A Matter Of Time

By: David W. Milliern

ISBN: 1-4107-5512-6 (e-book)
ISBN: 1-4107-5511-8 (Paperback)

This book is printed on acid free paper.

This edition was published by First Books, 2595 Vernal Pike, Bloomington, IN 47404-2782
Book cover created by Dawud A. Duncan, owner of D.A.D.'s art studio
Edited by Wesley W. McCulloch

1stBooks - rev. 6/19/03

This book is dedicated to
the Dark Lady, Lord Muck, and ytka.

"It amazes me, the will of instinct."

Table of Contents

Foreword

From ancient times, there have always been legions of theories describing the physical world. While our more modern theories have progressed beyond earth, wind, water, and fire into particle theory and beyond, there remain some of the same tests to be passed. A theory and its proofs should be able to to explain the phenomena, and do so without added magic, imaginary particles, or other additions for complete explanation.

This is not to say that the truth automatically triumphs at this point. As we know Copernicus' orbs explained very nicely when and where a variety of heavenly bodies were going to be at any given time, but were ultimately inaccurate. Copernican orbs were circular orbits. Present wisdom knows these orbits to be elliptical. Copernican circular orbs were accurate enough to be used by the earlier, less capable, NASA computers in their calculations of where to aim their probes, but as a total theory Copernican orbs were not the truthful paths taken by the planets. Thus, theories which explain a great deal may be inadequate.

This work is submitted with this knowledge; hoping it may clarify many quantum quandaries without additional magic, imaginary particles, or any other fudge factors. This work is by no means exhaustive, and subject to the same fallibility which time and testing may offer. Instead, it is hoped that these thoughtful postulates will inspire others more studied to see the music of physics as viewed by Mr. Milliern, rather than simply the musical score, with which they have become so familiar. As with any wonders of this powerful universe, it is assumed that the thoughts stimulated will produce new marvels, rather than the sinister which lurks in any quantum advance.

Whether or not Mr. Milliern's name may be spoken of in the same breath as the accepted all-stars of the academic past is irrelevant. What is more pressing is that Mr. Milliern's thoughts invite a serious audience. Had Albert Einstein's thoughts been given serious review when he was a postal lackey, the world would have developed quite differently. You are invited to be this audience for Mr. Milliern, and it is hoped you will find something here which tweaks your thoughts and postulates. We trust it is only a matter of time.

Abstract

The following theories deal ostensibly with the geometry of space and energy matter relations. The underlying theme is the universe's structure and its function. The author will go into some detail in explaining the current theories, and the suggested theories proposed, and how they effect and relate to one another. In some cases mathematics will be provided for those interested. Presently, we are in a somewhat stagnant period in physics, which has followed the revolution known as the beginning of modern physics. The author proposes redefining of the theory of fields and proposes unification and explanation of many theories that are currently well known. In doing so, the author hopes to stimulate subject of physics into the post modern.

The author will attempt to explain the relations of time to energy, and in some detail, what is going on in conversions of energy to matter. The author will also explain why energy does not have mass and matter does, and the idea that unites the two. In tying these together, the author will then show the relations of matter-time-energy-space in a unified paradigm of energy in its many forms.

The author will then lay the framework for a potential theory of everything. If the initial unified field theory is going in the proper direction, a theory of everything may be feasible. In laying such a framework, the author will go into some detail in the subject of waves. The purpose of this work is not to mathematically present a grand unified theory. Rather, it is to present a foundation for the GUT.

The GUT would not be complete without the unification of General Relativity, Quantum Mechanics, and Special Relativity. I will deal a little in chaos and a few other subjects, which are not my primary focus, so I will not dwell on them. We will begin with the simplest of these tasks…time. This work is mostly a continuation of the theory of light and space, which was contributed to by Issac Newton, Nikola Tesla and Albert Einstein.

Chapter 1

TIME

Ever since the dawn of man, we have pondered over the idea of time. Does it exist, and what is it? In the early days it was thought that time flowed in one direction like a river in either the x, y, or z (i.e., Cartesian dimensions). But as early scientists found, they could not go backward in time, or even slow it down by traveling in a particular direction (early Greek philosophy). Thus, it was thought that time was invariant (Galileo). Even the thought of time's invariance was done away with after the coming of Einstein as we will discuss later. Time is said to be a temporal dimension rather than a spatial dimension. We can physically traverse ("touch") spatial dimensions, and yet the so called temporal dimensions cannot be "touched", in the physical sense. Zeno proposed a set of paradoxes known as Zeno's paradoxes. They dealt with distance and time. The first paradox stated that by definition, traveling is to exist in every position, at some time, on a journey from "A" to "B." Given a point "A" is two feet from point "B," and an object moves an inch in a straight line between the two points, there is an inherent problem. The object must have existed in the half inch point at some time. If we go further to say that the object must have existed at the fourth of an inch point, we can cut the distance an infinite number of times. We can express this as: d/(2n)

Where d is some finite distance and n is some number approaching infinite. The implication of this paradox is that it shall take an infinite amount of time to travel an infinitesimal distance. Zeno's paradox is one that is solved via quantizations.

It has been argued that even though relative positions may change, there is no such thing as time. Let us first find out whether time exists at all. We will use a method called the "Flatlander idea." This idea was produced by Edwin A. Abbot, and mathematically defined using Georg Bernard Reimann's geometry, otherwise known as Reimann's geometry. This method is used in analyzing dimensions. After all, that is what time is. If we take a zero dimensional plane (a point) and fold it, we will have a one dimensional plane (a line). It may be difficult for one to see that a point, or zero dimensional plane, is perpendicular to some line. We shall express this mathematically, using vectors. We shall begin by changing some line "u," into vector "u." Assume vector "u" has a finite, non-zero length. If the cross product of two vectors equals zero, then the vectors are perpendicular. With little inspection, one could easily find that the determinant of the matrix (0,0) and (x,y), is 0ĭ+0ĵ. This means the vector (0,0), or our zero dimensional plane is perpendicular to vector u=(x,y).

Folding an x-dimensional plane (where x is some nonnegative real number) is the equivalent of simply adding another plane to the system. I will use the term "folding," because that is what is actually happening. If we continue and fold our one dimensional line, we arrive at a two dimensional plane, analogous to a sheet of paper. Holding a sheet of paper with the thin side facing you, fold it, and two dimensions is the result.

We can imagine people living in a two dimensional plane. The two dimensional world is known as "Flatland," hence the flatlander Idea. This will be the first of many thought experiments we will use. The people in flatland can only operate in two dimensions. They are oblivious to things in the higher dimensions, because in an x and y-dimensional space, there exists an infinite number of x-dimensional planes. An example of this is add another dimension to our two dimensional flatland, perpendicularly. It can be said that the two dimensional flatland has been folded. The result of which is a three dimensional space. The interesting thing is that a three dimensional being would be godlike to the two dimensional flatlander. For example, the three dimensional person would be able to flip over, and change the flatlander from being right-handed to left-handed, and vice versa. Yet the flatlander would be able to see an infinitesimally small portion of the three dimensional person at a single time. There is much to be inferred from this. What of time? If time is a dimension, it would be possible to show exactly what time is by using the flatlander idea. If axis lines are added perpendicularly to the three dimensional space, time will be more understandable as a dimension.

The first thing that we must do is examine all that we know about time. We know that if we walk in directions x, y, and z (plus or minus arbitrarily chosen), we will not travel into the future, slow or speed up time, or go to the past. We are temporarily neglecting special relativity for important reasons to be revealed later. For now we will follow closely to classical theories of time. We know that we do not head into the past as we go in a particular three dimensional direction, thus time does not flow like a river. Moreover, Stephen Hawking suggested that time travel is improbable. He said, "If time travel were possible, then why haven't we been taken hostage by terrorists from the future or visited by tourists from the future." Of course, that statement makes the assumption that future beings discover the technology to utilize such physical properties, if it were possible. Again, there are other paradoxes, such as the following: You are born by a certain pair of parents. You go back in time, and kill both of them. Thus you could not have gone back in time to kill them, because you were never born. Moreover, if you were never born, you could not have gone back in time to kill them, and then you would have been born.

These types of absurdities are chronic in dealing with the concept of time travel. We see from the previously stated, that time is flowing in all six directions of a three dimensional space, all at once. This makes sense since each newly added dimension is perpendicular to all of the previous dimensions. If we add the x axis to the y axis, you see that the y axis is perpendicular to the x axis. If you were to add the z axis, it would be perpendicular to both the x axis and y axis. With that known, we may fold our three dimensional region. What we have afterward is a three dimensional space, placed onto a flat two dimensional sheet. Higher dimensional visualization is by far the most difficult skill to attain in physics. Many mathematicians can describe higher dimensions to a point of fallacy in mathematical terms, but Einstein may have been the only person in history capable of beginning such an endeavor visually. What we should see in our minds is a picture in three dimensions placed on a two dimensional sheet with the ability to see everything that is possibly visible in three dimensions. If you want to get an idea of how such might look, there are many pieces of modern art that have been done between 1905 and 1950. In these works you can see both eyes of a person while looking at his/her profile.

For example, if we were to take a three dimensional picture of a man, and put all of the information on a two dimensional sheet, we would be able to see everything. By saying everything, it is meant that his outer appearance from all angles and his insides (heart, brain, veins, etc.) from all angles. So seeing the universe in four dimensions would be like seeing it on a two dimensional sheet from all angles at once. I have come to call this fourth dimension of space, the "Perpenulus." Let us say that there are two beings. One of which is a robber and he is a three dimensional being. We will call him "A1." "A2" is an everyday ordinary man (almost). The thing with A2 is that he is a being of four dimensions. So, when A1 attempts to rob A2, he had better think twice, because A2 not only sees the gun in A1's pocket, but sees his every move. The added piece of information now is that A1 will try to sneak up on A1 from behind. To no avail, A2 sees everything in front of him, behind him, and even objects not visible within the three dimensional perspective (like items within a closed box).

As wonderful as these findings are, this is not yet time. Something very important is missing. Suppose we took a picture of two blocks on a table, and later, without changing any arrangement, we took another picture. Now we take those two pictures and ask someone, "How much time has past?" Of course, the person would say, "I do not know." Thus, we have our first observation in process of finding the missing link. Now let us assume a second scenario: There exists a universe unlike ours. In this universe nothing moves. There are no radioactive elements, no stars, and no life. The question now is, "Does time exist in this other universe?" There is direct relationships between these scenarios. The answer to both is change! The two pictures may very well have been taken at two different times, however that is arbitrary at this point. The pictures may very well have been taken at the same time. There must be a change in at least one set of reference points. Thus, time does not exist in the theoretical universe in which we depicted. In both cases, there is an absence of change. This is the answer to our question, "What is missing from perpenulus that does not make it time?" Again, let us expand upon our dimensional analysis of the four previous dimensions by bending it. To do this, we take the three dimensions (appears to be flat in four dimensions) and perpenulus (the fourth dimension, which seems to be flat in conjunction with the first three) and fold the region. Because we have put the four dimensions into a two dimensional format, we can fold it into a seemingly three dimensional region (even though it is really five dimensions).

In folding the four dimensional region, we attain a five dimensional region unlike anything we could have expected. The five dimensional space consists of four dimensional perpenular frames lined up adjacently to each other in the fifth dimensional plane. If we look at perpenulus from a side view, we see one a dimensional line and if we bend it we will get a two dimensional plane. To give the reader a good visual image, stack paper into a consecutive pile of papers much like a cartoon flip book. Each sheet represents a perpenular frame, and the dimension by which the sheets progress in number is called the fifth dimension. What we have are frames of events (perpenulus), and a change in perpenular frames occurring in the fifth dimension. For reference we will call perpenulus the μ direction, and we will call the fifth dimension the Δ direction.

The problem we are faced with is the fact that the perpenular frames are pressed together, giving us no differentiation between frames. This problem is easily solved. We see that time is nothing more than the physical distance between two perpenular frames. Moreover, we know that distance is quantized, therefore time must be quantized. By making there an absolute minimum physical distance between frames, there will be an explicit differentiation between any number of frames.

Our next exploration in time will deal with Special Relativity, which was formulated by Albert Einstein, but first we will look at the founder of Special Relativity's math. His name is Heindrik Lorentz, and his solution was called the Lorentz Factor, which he found circa 1900. What he unknowingly found was that time was relative (among other things). It took Einstein to realize what Lorentz had done, because Lorentz didn't have the theoretical capacity to extract such unusual implications as Special Relativity. We will now explore Lorentz's mathematical work on Special Relativity:

1. $r^2=x^2+y^2+z^2=c^2+t^2$
2. $x^2+y^2+z^2-c^2t^2=x'^2+y'^2+z'^2-c^2t'^2$

3a. $x=a(x'+vt')$

3b. $x'=a(x-tv)$

4. $t'=((x/a)-x')/v$
5. $t'=((x/a)-ax+avt)/v$
6. $t=at'+(x/v)(a^2-1/a)$
7. $x^2+y^2+z^2-c^2t^2=a^2(x-vt)^2+y^2+z^2-c^2[at-(x/v)(a^2-1/a)]^2$

8. $x^2-c^2t^2=[a^2-((a^2-1)/a)(c^2/v^2)]x^2+2[(a^2-1)(c^2/v^2)-a^2]vxt-(c^2-v^2)a^2t^2$
9. $(a^2-1)(c^2/v^2)-a^2=0$
10. $a^2=1/(1-(v^2/c^2))$
11. $a=1/\sqrt{(1-(v^2/c^2))}$ (the Lorentz Factor noted (γ))
12. $(a^2-1)/a=(v^2/c^2)/\sqrt{(1-(v^2/c^2)}$

13a. $x=(x'+vt')/\sqrt{(1-(v^2/c^2))}$

13b. $x'=(x-vt)/\sqrt{(1-(v^2/c^2))}$

14.$y=y'$

15. $z'=z$

16a. $t=(t'+x'(v/c^2))/\sqrt{(1-(v^2/c^2))}$

16b. $t'=(t-x(v/c^2))/\sqrt{(1-(v^2/c^2))}$

Final: $t'=t\sqrt{(1-(v^2/c^2))}$

The resulting equation is called the time dilation equation. In essence it says that time is not invariant. Moreover, it varies with the difference in relative velocities in the observers. Although Lorentz had the answers right in front of him, it took Albert Einstein to explain the results (most of which we will look at in the following chapters). Einstein suggested that as an object moves through space at a faster, and faster relative speed, the time within the space of the object slows down. That is, as we from earth watch a spaceship pass the earth, we observe the clocks on the spaceship to run slower than our own. Einstein knew at this point that space-time was bendable (this will be better explained in General Relativity). Again, this proved that Galileo, and Newton were wrong about time being invariant. In fact, as Einstein extrapolated from the mathematics, the variation in time flow was greater as you approached the speed of light.

From a standpoint, we see that space tends to resist objects in different inertial reference frames (different relative velocities). We shall call this "Perpenular Friction," and as we go along, this concept will become more self-evident. Since space-time bends, we can say that there is some resistance between the object and space-time, which is why the term above has been called "friction." It is as if the universe wants everything to be in the same reference frame. Our results cause dragging of space-time. What is actually happening is the regions of perpenulus are bending. This explains why Einstein's statements on the chronology of events depend on the inertial reference frame as well as does the rate at which time flows. The question is, "What is happening to perpenulus to make such observations a reality?" It should be pointed out that each perpenular frame is an instantaneous picture of the universe. Just as a cartoon flipbook works. However, the universe resists objects in different inertial reference frames. The result is the dragging effect on all of the dimensions of space. Observer A is in relative motion with respect to observer B. Observer A believes himself to be at rest. Observer B believes himself to be at rest. Who is right?

Imagine a stack of perpenular frames, initially planck length (the smallest possible distance) from each other. When an object is in relative motion, perpenulus bends. What you would find is that the object in relative motion elapses less perpenular frames, from the viewpoint of the observer considered to be at rest. You will find that the observer (who considers himself to be moving) stretches out the duration of each particular perpenular sheet. The result being that the observer considered to be in motion travels through more space than time. Hence, time slows down. The observer in motion bends perpenulus such that the sheets become closer and closer until they are a melded together. At this point, when the sheets have melded together, one would be traveling at the speed of light. Thus, light travels through no time, rather space only. The photon's perpenular frames become the same frame, and thus they do not travel through time. As stated earlier, a change in perpenular frames must occur. It is important to take into account what photons experience as they travel between two points. If a photon could be asked, "how much time does it take to get from one end of the universe to the other?" The photon would say, "It took me no time to travel across the universe." A further pictorial representation would be to revert back to the stack of perpenular sheets. If we fold the stack of sheets at their midpoint, the upper half of the sheets will meld together and touch, whereas the lower half would remain at planck distance away from the next.

Let us now, more thoroughly analyze the perpenular frame diagram. We see the perpenular sheets from the side, and these appear to be a line. This will be the point of view we shall use to attack this problem. On each line there are points "A", "P" and "Q," where "A" and "Q" are ends of the line, and P is some arbitrary point betwixt A and Q. As some object Z in relative motion speeds up, the angle θ (which lies between the incident line in the delta direction and the segment P to Q) decreases. When the speed of light is reached, the segment P to Q overlap segment P to Q in the next frame when θ=0° with respect to the incident line. If you think about an object existing in multiple frames, and is stationary, the object is said to be traveling at the speed of light (denoted by c). There is no preferred inertial reference frame (Principle of Special Relativity). Two objects at rest must be at some constant velocity, and zero meter per second is only relative. Both must be at a constant velocity, and the only constant is the speed of light. This implies all objects travel at the speed of light in order to exist. Moreover, matter may only exist if it is traveling at the speed of light. This is the reason for c^2 popping up in the mathematics so much.

It is now arguable that time is not a fourth "temporal" dimension, rather the it's the physical change of μ (perpenular frames) in the Δ (fifth dimensional direction). Time is possibly the interaction between the fourth and fifth dimensions as opposed to the currently accepted idea of being a fourth dimension all of its own. This is slightly more obvious when one realizes that the fourth dimension is nothing more than instantaneous pictures of the universe. These pictures must change in the fifth dimensional direction, and the interaction is analogous to a cartoon flipbook. In this scenario, we also see that planck time is the physical distance between perpenular frames, which makes more sense of the quantization of time. It is now somewhat obvious that planck time, and planck distance are the same measurement. The only difference is that the measurements are in different dimensional planes. As will be explained shortly, perpenulus is not only crucial in the explanation for General, and Special Relativity, but it defines physical behaviors. Furthermore, the proof of its existence is but a few chapters away. As in all physics, we must first make our observations, suggest governing rules, and conclude with mathematics.

In knowing what time is, we see why Zeno originally had such a paradox. The thought of quantization only arose in the 1900's. Perpenulus is a macrological phenomenon, and will remain so even at the end of these works. From Zeno we were left with the remnants of his paradox, to Galileo, who thought time to be invariant. This has been completely washed away in Einstein's work, and now we have a clean slate "needing" to be filled. That slate has been filled in this chapter with an explanation of what time is, and the nature of it. The structure of the universe maybe a multidimensional space-time including variance in observation with speed, and frame to frame differentiation so long as the speed of light is not achieved. At the speed of light, the perpenular frames meld together in that region of space. For example, as a photon travels at the speed of light, there is no differentiation in the perpenular frames in the location of the photon. Thus, the photon does not experience any change in time. Using dimensional theory, we have already laid the foundation for the next theories to come. Perpenular Friction happens to be a determining factor in much that is to come and it is important to grasp the concept early on in this book.

Chapter 2

Waves

The world of physics is constantly hunting for answers. These days it would seem the hunt is fully directed toward the Theory of Everything. The idea involves compressing all knowledge into a single equation. The equation would describe all of history, present events, and future events, behavior of everything via laws of physics. The universe (our closed system) could possibly be described by one equation. At the present time, we are nowhere near arriving at that equation, should it even exist. In fact, many physicists doubt that there is a single equation of such magnitude. One must question the fact that if there is an equation that simplifies everything, is it possible to extrapolate concepts in the opposing direction. For example, what color the sky is on earth? The equation for everything would, in theory, contain all of the information of the universe, or at least the information could be inferred from the equation. It is one thing to derive an equation that describes everything, but it is a completely different scenario when one is trying to infer what the universe will be like by simply inspecting an equation. The question is a simple one. It is a question that will only be answered if we first work backwards to get the equation, and then try to achieve a universal model from it. The author's thoughts are that there is an equation of everything, and it is probably something to the effect of A=A, where A is an arbitrary variable. Assuredly, such an equation will not tell me the color of a particular rose, as well as the physical laws that govern the universe.

Some believe there is an astounding possibility that there is a single mathematical statement that encompasses our closed system. Moreover, as with most mathematical systems, there are constants, a set of axioms and guidelines that the universe is subject to. Initial conditions are most likely unattainable. It is impossible to fully know the state of the universe, at any one point in time. As you will read in chapter five, energy is a condition which cannot be completely certain. The author has developed a few simple principles, and some complex principles to describe the universe, which will serve as guidelines for the equation of everything.

We shall begin with the single law that governs the universe and all that exists within it. This is the law from which all the laws of physics have been derived. This law is the "Law of Fundamentals." The Law of Fundamentals has two subsets:

1. For all things that exist, there is an equal and opposite.
    a. A+(-A)=0
    b. A=A
2. For all fundamentals in nature, equilibrium is the perpetual desire of nature.

The problem most people will have with the Law of Fundamentals is deciding what a fundamental is. Immediately people see quarks, and anti-quarks, and believe these particles to be fundamentals. The problem is not that quarks are, or are not fundamentals. This is not yet known. The problem is that fundamentals can construct the same structures in nature, and be on opposing sides of the spectrum. An example is to say that a proton is the fundamental building block, because there is an anti-proton. This is surely false, because that there are quarks that are equal, and opposite in nature. The only thing that showing a proton has an equal, and opposite does is prove the proton is composed of true fundamentals. This proves that within the realm of our closed system, the proton exists by definition. A true fundamental is defined to be the smallest indivisible pair of equal and opposites. This goes for all things that exist; an example being Newton's Third Law.

When one mass collides with another, they both experience the same force as illustrated by Sir Issac Newton's Third Law of Motion:

$$F_{ab} = -F_{ba}$$

These are two fundamentals which are equivalent and opposite in nature. Just as if we look at a beam of light being reflected off of a mirror, the angle in equals the angle out.

If we take this idea even further we find that the true fundamental "behavior" of the universe is to be a wave. Waves are inherent in the previous ideas, so let us continue. A boy throws a ball up in the air. What we find is that the time it takes for the ball to reach its maximum altitude, equals the amount of time it will take for the ball to fall to the initial starting position. We also see that other things in nature tend to be symmetrical at simpler complexities (which we will talk about momentarily). In fact if we were to look at a sine function's graph, there is something interesting about the are under the function's curve.

$$\text{From 0 to } 2\pi \int \sin(x)=0$$

By looking at the first period of the sine function we see that it is wave nature to have an area of 0 after each respective period.

Some manipulation may need to be done as will be written about later.

This now brings us to the Universal Principle of Waves. This principle states that all behavior within our closed system behaves like a wave and characteristics can be attributed to a wave. This principle can be elaborated with the following examples: A ball roll down a ramp. There is more to it than meets the eye. I see the light reflects off the ball (equal angle in and angle out), the light "waves" travel through space, the sound waves (which are created from the vibrations of atoms) travel to my ears, the ball spins which is an oscillation, while it has some DeBroglie wave association with it. The atoms vibrate, because they have some temperature, the electrons have some wave probability in their position, and so on. All that has been mentioned in this example are waves. Moreover, everything in the universe is a wave, and all that occurs within this closed system can be attributed to waves. To ignore the previous scenario is to ignore the truth, which is reality. Absolutely everything that resides in our closed system is a wave. What this suggests is that the algorithm of everything should have a wave function of some sort. The following is the standard form of a wave function:

$$\partial^2 y(x,t)/(\partial x^2)=\partial^2 y(x,t)/\{\partial t^2(v^2)\}$$

This equation only has usefulness to us as a format of wave equations, so we know preliminarily what we may be looking for. The nice thing about wave equations is that the wave need not be sinusoidal. This means that the equation can be used in dealing with a wider range of waves like electromagnetic waves. As you may have noticed the equation has both x, and y, as coordinates, and t for time. What does it all mean? You use what is called wave propagation to find (x,y) at some time, according to the wave function. In addition, DeBroglie made the discovery that all objects have a wavelength which can be found by the following equation;

$$\lambda=h/p=h/(mv)$$

If these theories are correct so far, then all things behave as a is wave. Through further investigation it will be found that all things have wave characteristics, because they are waves. There are other things to be considered. Waves are nothing more than patterns. In fact, sequences of numbers can be used to describe a wave. What we have arrived at is one of the most controversial subjects in philosophy, determinism. The interesting thing is that all sequences of numbers represent some pattern. Hence randomness is no longer a reality. The only challenge the author will make to the philosopher is to assume that you as a person have some free will. If this is true to even a small degree, and you accidentally broke your antique vase, change the out come after it has happens. If one had free will, and his/her will being that they wanted to preserve the vase, he/she will have the ability to change unwanted (unwilled) outcomes. Thus one's free will has been infringed upon.

Finally I must make known that all waves are the same wave. They may be skewed, off beat, different amplitudes and so on, but the key thing is when small constants are added to any wave equation, they can be made exactly like another. It may seem like common sense that one could add constants, multiply amplitudes of waves, change frequencies, and so on, but the concept is very important to realize the relationships that exist in the universe. Once it is realized that all waves are the same wave with some addition made to the basic sine wave function, it is possible to see how everything unifies to one idea (i.e. matter, motion, energy, etc.). The topic of interest then is how a single wave function manifests itself into such macrocosm, and why it does so.

When using the terms complexity, order and entropy, concepts can be confused, and even misused. This is probably, because there is a concept that is currently missing from physics today. I call it Milliern's Principle, which deal's with the complexity of systems. What is complexity? Complexity is accepted by the physics community to be the level of entropy in a system. Complexity shall have a new meaning in this book. Complexity, as I have defined it, is the capacity of potential break down within a system, or potential for a system's failure. The potential for breakdown can be represented as probability, but we will not explore the actual mathematical representation. Such an equation is far beyond the scope of this book's purpose. A direct ramification of a breakdown is that the system shifts very rapidly toward equilibrium. The state of equilibrium would be when the system has decreased it's complexity to the point where the system is no longer unstable. Complexity, both increases and decreases as disorder increases. There can be an increase in entropy and a decrease in complexity. An example would be in the formation of galaxies. As currently defined by the physics community, the gases that are coming together to form a galaxy, are increasing the order of the region. Furthermore, no other regions of space are compensating with disorder for such local increases of order. With the proposed definition, there is no implication that the order is actually increasing, because complexity as a term is no longer interchangeable with the term "disorder." As a matter of fact, disorder can not occur without some change in complexity. As time goes along, disorder or entropy, increases. As the disorder increases, complexity increases in order to achieve an even higher level of disorder. The Second Law of Thermodynamics holds true under any, and all circumstances. The complexity of a system merely increases so that the disorder of a system can occur more rapidly. As that complexity increases, the potential for breakdown increases. The addition of parts to a system is a major factor in the increase of complexity. By adding parts to a smaller system within the whole, we increase disorder. There is no law of conservation of complexity in the universe, therefore, it is possible to look at a small system as being closed in terms of its complexity, but not it's entropy. In adding these parts to the system, they either add to the micrological system's complexity, or make no change. The negative change in complexity, if there is one, comes from the macrological system. Thermodynamics also, suggests this by saying; there may be an increase in order in a smaller isolated system, if the whole system takes account for this by increasing overall entropy.

In dealing with complex systems we have inescapably fallen into the new physics known as chaos. Chaos was distinctly identified by a meteorologist by the name of Lorenz. He first proposed what is called the Butterfly Effect. The Butterfly

Effect asks, "If a butterfly flaps it's wings in China, can this disturb weather patterns on the other side of the world." As Lorenz was playing with a thought experiment and calculation machine, he found great variances in his predictions in weather patterns after two, or three days. He even produced a famous experiment called the Lorenz Wheel. The wheel had eight buckets with uniformly distributed holes, and a water source that flowed at a constant rate directly over the top of the wheel. What Lorenz found was a completely erratic pattern of wheel rotation. The wheel would rotate in counter clockwise and clockwise directions without pattern. As a matter of fact, all of the patterns in complex systems are absolutely unpredictable. This goes against all that physicists have believed for centuries. We believed that the universe we live in is predictable if all information is given. Laplace expressed pre-chaos theory in saying, "if there were a mind that could calculate all the information that exists, we would be able to predict everything exactly." As it has come to be, the world appears to be dictated by sets of non-linear mathematics. Much against Laplace's suggestions and computers cannot determine stock exchange patterns. Attractors for individual systems have been developed, but appear almost meaningless outside of that specific system. Attractors are simply mathematical expressions of chaotic systems. By looking over a few of these attractors, one can develop an appreciation of how difficult these systems are to understand, and express mathematically.

The attractors are fairly simple. For example the Lorenz attractor is as follows;

A. dx/dt=a(y-x)
B. dy/dt=x(b-z)-y
C. dz/dt=xy-cz

Lorenz's attractor is used for describing two gaseous bodies of different temperatures and their behavioral motion respectively. Another famous attractor is Navier-Stokes;

A. dx/dt=p(x-y)
B. dy/dt=Rx-y-xz
C. dz/dt=xy-By

Where P=10, R=28 and B=(8/3).

Rossler's attractor is also important in chaotic systems;

A. dx/dt=-y-z
B. dy/dt=x+ay
C. dz/dt=b+z(x-c)

Where a=.2, b=.2 and c=5.7.

And finally is the Duffing attractor;

A. dx/dt=y

B. $dy/dt=x-x^3-ay+b\cos(wt)$
Where a=.25, b=.3 and w=1.

Each of the attractors describe their own types of system, but why these particular equations are as they are is unknown. Why exactly these seemingly meaningless patterns occur is unknown. Moreover, the shear capricious nature of these systems is now one of the world's largest research topics. Chaos is an endless topic that brings all of the mathematicians, and natural scientists together.

Now that we have a definition for complexity, and a miniscule understanding of chaos, let us indulge in Milliern's Principle. It states, that as the disorder of a system increases, so does complexity until an eventual break down occurs. At this breakdown, the system attempts to attain equilibrium (stability). Under the initial conditions of a particular system, the system can only achieve a certain maximum complexity until the initial parameters are improved. Disorder is a term that comes from Thermodynamics. The second Law states that our system will never go from a state of low organization to higher organization. This is called entropy. Otherwise stated, entropy is a measure of disorder.

The equation for the potential for a system to break down is;

$$\Delta G = B - U_p$$

Where G is Gibbs free energy, and B is the energy needed to cause the system to breakdown and U_p is the potential energy that can possible be released in the system. Why does this look so familiar? It may look familiar probably, because it is a widely used equation in chemistry, and some physics, known as the Gibbs free energy equation. It has not been confirmed experimentally, but the equation seems to have worked in many cases in the math. B, the energy needed for breakdown, can be calculated by using triangles where the legs are the height of the system and the diagonal of the square base. The thought experiment the author used involved wooden blocks. As blocks are stacked higher and higher, less torque ($T=Fd\cos\theta$) would be required to knock the system over. Moreover, there would be less work ($W=Fd\cos\theta$) needed to break down the system, assuming the parameters are constant. The parameters in this case would be the area of the base. One would have to agree that with any fixed set of parameters that there is a maximum complexity to the system. This is so because there will come a point where the force needed to break this system down, would be equivalent to the slowest moving gaseous particle in a room, at near zero temperatures. Under no conditions would the complexity be exempt to break down causes. Even if you had a nearly perfect closed system, the mere background radiation of the universe may upset the tiniest of subatomic particles, and the wooden block system will come crashing down.

A good example of mine is nitroglycerine, $C_3H_5(ONO_2)_3$.This is the compound that is absorbed on diatomaceous earth and was called dynamite by its founder, Alfred Nobel. When nitroglycerine is slightly jostled, the bonds break and release large quantities of energy leaving $\rightarrow 6N_{2(g)} + 12CO_{2(g)} + 10H_2O_{(g)} + O_{2(g)}$. The compound is saturated with high potential bonds. The same reason we use tetra methyl butane for cars (C_8H_{18}), and other octanes, because this 100 octane is saturated with a large amount of potential energy in the bonds.

If we were to "randomly" toss blocks onto a pile, they would gain complexity, and sooner, or later fall down. This is a collaboration with the previous idea of adding parts. It is much like the stock market, legal systems and especially evident in physics.

Let us suppose a construction company builds a tall building and it continually gets higher. The disorder is increasing because energy is being lost (to useless heat energy), and the arrangements of matter are changing, thus becoming more defiled (less ordered than originally). The other thing is that the complexity is increasing as the building gets taller. If the building fell over, there would be a drastic increase in entropy within the system in just a few seconds. As you see entropy only increases, never decreases. In a completely closed system like our universe, it is possible to have increased regions of order, but collectively the system always lessens the order for which it has. It can be looked at as the increase of the overall average entropy of the system. Complexity on the other hand, does both increases and decreases. The increases of complexity occur gradually (compared to the breakdown in most cases), and can break down equivalently, but more often then not, equilibrium is achieved with in a short period. It seems as though the universe doesn't have such laws of complexity because of extemporaneous choices in rules. The universe's goal is to achieve complete entropy in a desirably minimal amount of time. The continuous increase of entropy is less desirable than to have times of spontaneous and releases of energy, or instantaneous failure of a system.

We get a skewed bell shaped curve. The curve in general is a constant format, which implies that there is some kind of numerical constant in dealing with complexities of systems.

What we see in a particular systems growth is an exponential curve, as well as many other expressions of exponential growth and decay. The natural number e, seems to dictate the curve. Why this is the case, I am not sure. I do suspect that there is a relation between e and the golden ratio (which defines the stability and complexity of a system). Before we go any further into the correlations of e and Φ (the golden ratio), let us explore what exactly the golden ratio is and why it is important.

The golden ratio was originally noticed by the Greeks and they realized it had a great importance in nature but they weren't sure why. What they did know was $\Phi=(1+\sqrt{5})/2$. Which is approximately 1.618033989. Greeks found this answer by using the quadratic equation:

$$x=(-b+\sqrt{b^2-4ac})/2a$$

Where they took a=1, b=-1 and c=-1. And if you have a great deal of time on your hands, mathematicians also derived the number from the following:

$$1+(1/(1+(1/(1+(1/(1+(1\ldots= \Phi$$

It is interesting how complexity deals with the addition of parts to a subsystem and the previous derivation deals with an infinite addition of parts. Later a man discovered that a recursive sequence, which is now called Fibonacci's numbers, has a limit of Φ when divided by the second number. For a little background, a recursive sequence is dependent upon the previous n numbers (where n is some real number). The recursive sequence is: $f(n+1)=f(n-1)+f(n)$ where $f(1):=1$and $f(2):=1$ and $n\geq 2$

The sequence goes like this:{1,2,3,5,8,13,21,34,55,89,144,233...}. When the sequence divided by f(n) such that we have the expression:

$$[f(n-1)+f(n)]/f(n)=\Phi$$

This number Φ and ratio Φ:1 are ubiquitous in the world, and maybe even more preeminent than π. Φ appears in art, structures, exponential functions, and as we'll see in complexity. So what exactly is Φ and the golden ratio? The problem is that no one knows for sure. The author's suspicion is that the numbers can be derived from one another and Φ is the ratio of parameters to a system's state that defines structural integrity of non-static systems. What we do know is that, as anything in nature becomes more complex the more obvious Φ becomes. Beauty has also been related to the golden ratio, and the complexity of various objects. The golden ratio can be described by two particularly well known geometric representations. They are the golden spiral and the golden triangle.

The golden spiral starts with rectangle where one of the sides are Φ times longer than the other. Then another rectangle is produced with the same ratio of sides as initially. Then this is continued, and we connect the arcs of the rectangles, which gives us a spiral unto infinite. The other interesting thing is that the Fibonacci sequence occurs often in nature. For example, humans have 2 hands with five fingers consisting of three parts separated by two knuckles on each finger. Also, if you ever look at a sunflower, you will notice that the number of seeds on the nth-circumference is $f(n)=f(n-2)+f(n-1)$. As we look into the nature of π and Φ, we see that π is the static structural constant thus defining the structural integrity of the static universe whereas Φ is the structural constant of the changing universe.

Life, and the actions resulting are naturally occurring. However, I will for the sake of argument, state for you a few more examples. The example of the building could have been involving a tree instead. We will look at radioactivity in our next example. It is said that the desired proton to neutron ratio in an atom is 1:1. The ratio of neutrons to protons is most stable at one to one and least complex. Hence, it is completely stable in terms of radioactivity. As protons are added the parts to the system the complexity increases. There are two critical points in the growth of complexity in systems. The first is at which point the complexity is at its lowest potential for which breakdown will occur. The other is the maximum complexity possible for the system to reach, given a specific set of initial conditions. Nature continuously attempts to surpass previously achieved levels of complexity to achieve even greater exponential increases of disorder. This is done by increasing the quality of the parameters, and in doing so, increasing the maximum achievable complexity attainable.

All things in economics, law and biology follow this behavior. Absolutely everything seems to be governed by it. Regardless of debates over evolution, it agrees with the definition of complexity, as stated in this book. Once breakdown occurs, nature tends to adjust the parameters to better suit a higher complexity. This is Milliern's Principle.

When we look at complexity and the constants natural e and Φ in retrospect, we see that Φ defines the structural integrity of non-static and e defines the graphical changes in complexity for an individual system under its own circumstances. We see this in population growth or any growth and decay for that matter, discharge of a capacitor and so on. The mystery of growth, and decay rates has now become a prime interest of natural scientists. The question that comes to mind is, "why is natural e 2.718, and not another number?" What is the significance these constants, such that they appear in so many place in the universe? All that can be said is that they are very much related to π. One interesting thing is the following equation:

$$e^{\pi\iota}+1=0$$

This equation contains five of the most important constants in mathematics, and yet the golden ration is not present.

The other thing one must take into account is the various types of wave behavior, and they are called the Rules of Wave Mechanics. Everything is a wave, and so we should define the nature of their various behaviors. There are three types of wave behavior in our system and they are equilibrial, assertive and paradoxal. There can be wave behavior which falls under more than one of these types. Each of these wave behaviors oscillates in accordance to eventually reach equilibrium, thus satisfying the second stipulation of the Law of Fundamentals.

Assertive behavior is when there are two extremes and no medium between. Some function, f(x), is such that it oscillates, not gradually, but occurs in an instant. An example of this is easily displayed in everyday life. If some real person named Bill has a book in his hand at time x=1 (where time only changes in nonnegative integers) and at x=2 he doesn't. What we see is an instantaneous change between two extremes. No equilibrium is achieved. In fact, there is no real or existing line of equilibrium. Moreover, the only way to achieve equilibrium is to unify the two points (in this case pertaining to space).

Equlibrial behavior is when there two extremes and there is a medium. In this behavior there are two classifications; transversal and longitudinal. In both cases a subject crosses though a non-extreme region to reach an opposing extreme. Equilibrium may be reached in this type of behavior (i.e., decreases in complexity). Oscillations consist of gradual changes over time. An example of this is a pendulum. The particle we look at swing past the point of equilibrium and continue to the opposing extreme. If other wave functions are added, equilibrium in this system can be obtained (i.e., force: equal and opposite wave function). Of course, the cost of putting one system into equilibrium, is moving other systems out of equilibrium. Thus, the disorder increases.

Paradoxal behavior is when the subject exists at both extremes at the same time. No change in time is necessary for the subject to be at an opposing extreme. There is a medium, as well as a point of equilibrium existing. An example of this is a charged particle in quantum mechanics. In this case the particle is attracted by a field and is observed in two or more places at once.

Now we come to an awe inspiring relic of physics. This relic is what we know to be a force. The problem with the idea is we can describe forces to a very high degree, and yet no one knows what gravity is, or what causes it! The author believes that he has come to the understanding of what a force actually is. A force should be defined as compatibility, or incompatibility of wave functions. In later chapters we will systematically go through the current scientific data that support this finding. Moreover, we will derive even more from this definition. At the center MCG-6-30-15 there is a super massive black hole, about the size of Mars' orbit, spinning at nearly the speed of light, and is producing much force, and is dragging around space and time. Much in accordance, a much overlooked implication of Einstein's theory of General Relativity is the suggestion of space-time being dragged around by rotating objects. This is very important to the author's theory's confirmation, because the classical idea of force is known as a curvature in space time. If compatibilities, and incompatibilities of wave functions produce curvatures of space, this can very well be found through simple experimentation.

Dr. Eugene Podklentov while, at Finland's Tampere University, stumbled across something big. In experimenting with super conductors, he spun a disk that was cooled by liquid nitrogen, and spun by magnets. It so happened that he realized something was floating over top of the disk. Later, he repeated the experiment and slightly reduced the amount of pull the earth's gravity had on the mass. The author does not believe this to be anti-gravity, as will be explained in further detail later, because gravity is set up such that it cannot have a repulsion effect. The important thing here is that a force was produced via the compatible/incompatible of waves. The other thing to be pointed out is that the mass should have an attraction toward the disk. This implies one of two things: Either there is another force being produced, which counter acts the attraction. There maybe some sort of rudimentary right hand rule that we have yet to exploited, but that is somewhat unlikely.

The next topic of interest is of equilibrium and constant states. We shall call this the Boltzmann-Lorenz Law. This law states that no state of equilibrium, or any constant state will remain in such a state for an infinite amount of time. The universe is set up in an unusual manner. Although the most desired state of any system is equilibrium, the universe disallows systems within the universe to maintain this particular state. If one thinks about it long enough, equilibrium is a very complex state, in which any miniscule disturbance will compromise the state, because continuous states are in equilibrium in some way. This also goes for any particular continuous state. In short, one could say that the universe will never become stagnant. An observance of this is in a completely stable Helium atom. The Helium atom does not chemically react with anything, and the first ionization energy is larger than any other atom's first ionization energy. However, the fact that protons decay over long periods of time gives way to a new atom. This is because the atom is defined by the number of protons in it's nucleus. When the proton decays, the atom will then be a hydrogen atom, and will be susceptible to reactions. There are other ways to go about lessening the stability but this is one of the simpler means. The universe is built on the concept that nothing can stop it from the oscillation that will occur between existence, and nonexistence.

Nonexistence refers to pre-Big Bang when there was no matter, and energy. Moreover, there was no space-time in which anything could exist. The universe as we know it was nothing more than a point, and remember that points are theoretical, taking up no space._The most profound thought is the fact that all states change in the universe, and change comes at the cost of increased entropy. Furthermore, any change in entropy calls for a change in complexity, which increases until a breakdown occurs, thus resulting in a lower level of complexity.

One very important system involves nuclear physics/chemistry and that system is the periodic table. The author has found through a little investigation that the periodic table can be represented on a graph in a wave formation. In this format, the complexity is designated by the vertical axis, and number of protons designated by the horizontal axis. The wave formation is similar to the following format;

$$|\cos x| = y$$

The only difference is that the amplitude of the wave increases exponentially with every cycle. For example, the first cycle's peak is hydrogen and then comes to the horizontal axis with helium (the only atom to touch the axis. There is a threshold line parallel to the horizontal axis for which radioactive decay occurs. At this maximum lower bound, atoms are unstable. This pattern is followed over and over, again while the closest atoms to the horizontal axis are the noble gases.

Finally, in this endeavor into complexity, we see how it is related to the golden ratio and the importance of Φ. As I have stated so many times before, the non-static structural integrity is based on this number, but for intents and purposes we will look at only the golden spiral. Up to now, not much has been written about the golden spiral, but the golden spiral is the very key to the complexity of systems. The entire goal of systems in our universe is to achieve a greater proportions without the large potential for breakdown. In each system, after a system breaks down, or even during the breakdown of a system, the parameters adapt to meet the requirements to accommodate a higher level of potential. For this we shall go back to the wooden block analogy, and randomly toss the blocks onto the top of the pile. The blocks system will have points of breakdown but in these breakdowns the base's area becomes greater. Hence, it better suites a higher mountain of blocks with each and every addition of a block. This relates to the golden spiral. If we take the golden spiral to be three dimensional, we see that as the height of the spiral increases, the bases also increases. If we extend the spiral upward, and downward at the same rate radially, the cornucopia retains perfect stability. If the cornucopia is sitting in a system with some gravity, the cornucopia could increase it's height infinitely, and never risk falling over. Moreover, the catch has always been that there is a maximum potential for a system's complexity assuming permanent parameters. It is when those parameters become constant that the system will go to equilibrium. Knowing that nature is designed by this golden spiral, we also see that is we do not change the base, but we increase the height of the spiral as there is a point of cancellation such that the spiral cannot wind upward any more.

In this chapter we covered many topics that had almost nothing to do with each other, except that they all dealt with waves. They all dealt with waves. We talked about the importance of π and Φ and their meaning with respect to the design of the universe. Complexity is defined by the golden ratio, which does not permit the universe to become dormant on a static state. As Milliern's principle has stated, the increase in entropy causes an increase in complexity, until breakdown occurs, so long as the parameters are constant. Systems with constant parameters have maximum levels of complexity. We know that nature has few, if any cases, in which the parameters do not change.

Another key point of this chapter is that all things in the universe that exist are waves. Existence is defined by the Law of Fundamentals, and it's two parts. The first is that for all that exists there is an equal and opposite. The second is that all fundamentals in nature have a desire to achieve equilibrium. The second stipulation leads into the definition of a force, being the compatibility or incompatibility of wave functions. With these waves comes their characteristic types of behavior; assertive, equilibrial and paradoxical. The universe goes through such tedious measures to achieve equilibrium. If this equilibrium is even possible, the state can not remain so for an infinite amount of time. That is the Boltzmann-Lorenz Law. Through all of these findings, and works the goal of an equation for everything is something of an elusive mythical creature which may soon come to be a reality. The place to begin looking is, by looking for the underlying rules for everything.

Chapter 3

Energy and Deftonic Theory

In this chapter we will formally introduce the major theory of this book. It is called the Deftonic Theory. Deftonic Theory involves fundamental dimensional theory, and quantum mechanics as a unified entity. The Deftonic Theory involves exclusively with matter/energy, and the space-time it occupies. Deftonic Theory actually includes Chapter One, but to better understand the theory, we shall begin with modern physics. Modern physics began in the early twentieth century. Therefore, we shall begin with the father of modern physics, Albert Einstein.

In the early 1900's Einstein completed and published his Theory of Special Relativity. One of his immense breakthroughs involved energy and matter. Let us go down the line, and give a complete overview of Special Relativity before diving in. Special Relativity was composed of two postulates:

1. There is no preferred inertial reference frame.
2. The speed of light is constant to all observers.

This is probably the most well known of all scientific discoveries, not only because he was right in nearly all of his works, but also because the greatest minds could not understand his theories. His other great discoveries were that time is relative, as well as the behavior of mass, and length of objects in relative motion (see chapter one for the mathematics). Finally, the idea that mass, and energy are some how related was shocking enough, but Einstein showed they are actually different forms of the same thing. You may be familiar with $E=mc^2$, but this is not the complete equation. The actual finding is as follows:

3. $E^2=p^2c^2+m^2c^4$

This may not look like the true equation, so we will work backwards to prove it.

4. $E=\sqrt{}(p^2c^2+m^2c^4)$
5. $E=\sqrt{}((p^2m^2c^4)/(m^2c^2)+(m^2c^4/m^2c^4))$
6. $E=mc^2\sqrt{}(p^2/(m^2c^2)+1)$
7. $E\approx(p^2/(2m^2c^2))+1$ (we can do this via Taylor series)
8. $E\approx((mc^2p^2)/(2mc^2))+mc^2$
9. $E\approx((mc^2m^2v^2)/(2mc^2))+mc^2$
10. $E\approx(mv^2/2)+mc^2$
11. $E\approx KE+mc^2$

Where KE is kinetic energy? Again, we have a mathematical observance, which has been shown in experiments. We are left with a profile of observations, but no knowledge of what energy is exactly.

So what is energy? It is a one dimensional string. Now you are probably wondering how such a conclusion was arrived at. The author took the idea of energy and all that is known about it (physically and mathematically). Energy is present but is not seen, it is quantized, and it is matter in another form. There are three particular quantities in nature that are quantized and they are distance, time and energy. As recondite as it was, it was arrived at that time is actually a physical distance between perpenular frames. Would it be a completely haphazard assumption to think that energy may be a physical distance? In this way, energy is merely a one-dimensional length in space with no width. Everything thus far checks out as we look at the facts. We cannot see energy. If energy were one dimensional, then we would not be able to see it (lines are impossible to see in two dimensions and higher). Also, other physicists have, to some degree, arrived at the same conclusion. Now all that is needed is validation via indirect proof, or supporting evidence.

If we look back to the Lorentz transformation in Chapter One, it is shown that mass increases with speed:

$$m_0=m'/\gamma$$

Where m' is the rest mass, m_0 is the relativistic mass, and γ is the Lorentz Factor. We can show mass does increase with speed by doing the following math:

1. $\lim v\to c\ m'/\gamma$
2. $m'/\lim v\to c\ \gamma$ (via limit theorem)
3. $m'/0=\infty$

Therefore, as the speed goes to the speed of light, the mass increases infinitely.

Thus we are left in the most perplexing of situations. We must break through to find what exactly gravity is in order to find the correlation between matter and energy. Furthermore, we must do it with the knowledge we have arrived at so far. This may seem out of the realm of possibility. However, we have the appropriate tools. By starting with the definition of a force, we can find the innate suggestion that gravity is the compatibility in a particular pair of wave functions between matter particles. If we continue on, using laws of generality, we find that there is only one way that "n" particles could all have opposing wave behavior. This is made possible if, and only if each particle containing both opposing wave functions, positive and negative, at once. In theory, if every piece of matter could contain two wave functions, then all pieces of matter would be attracted to one another. Of course, there is a minor proviso, so that all pieces of matter do not repel one another. It is most likely that there are two wave functions to be dealt with on the subject of gravity. If the converse were true, and there are an approaching infinite number of wave functions, then every particle in nature would have to assume different behaviors. This is very unlikely, because waves are either in the positive or negative side of the line of equilibrium. Matter has never been observed to have two masses repel. Therefore, we shall assume that there are two wave functions in every particle.

The result we come to is that these strings of energy are actually spinning in space. Again, in our thought experiment, allow us to curve the spinning string into a "U" shape. At the distinct ends of the string, one will have a clockwise spin and the other a counterclockwise spin (arbitrarily). It only matters that you see both ends of the string rotate differently when bent, whereas the string as an entirety is in a uniform spin. Because the two ends are in an opposing rotation, by definition, a force is present on the opposing sides. The opposing sides themselves experience the force. If we bring into the experiment another particle composed of bent string, there will be a force present upon the two strings. We call this a gravitational field. This menagerie of rules in Chapter Two is actually of absolute importance. This will be evident in the more so complex ideas.

That is gravity in a nutshell. We only need to explore energy-matter relations further, and decide if this projected theory of gravity is indeed factual. The next step is to tie off any loose ends with the initial theory (no pun intended). Strings in the form of quarks, as a ramification of the previous, will be looped (and "sealed") strings of energy. This provides us with a steady system of mechanics for hadrons. Reaching back to relativity, we are faced with the greatest question of, "Why does mass increase with speed?" Our answer is more than conceivable. Thus far we know that space-time resists different inertial reference frames (perpenular sheets). Keep in mind that in order for an object's mass to increase, there must be a change in inertial reference frames, which requires energy. This energy is converted (from whatever form) to kinetic energy.

a. $\rho=mv$
b. $\rho=\int mv\ dv$
c. $KE=mv^2/2$

The kinetic energy does not leave the system. Because of this perpenular friction, the kinetic energy put into change the inertial reference frame is now being bent proportionally to the velocity. As we have said before, with the bending of energy strings comes gravity. With that known, it is more than conspicuous that as an object speeds up, there is an increase in energy within that region of space. Moreover, the bends in the "loose" energy increases giving us a greatly increased mass with respect to the speed.

Another "juicy" subject is on the matters of low mass particles, such as photons. Photons, in particular, demonstrate both wave like behavior and act as a particle. This can be perplexing without the knowledge we have just discovered. Instantaneously, we see that photons have some relativistic mass. Thus, we know it is bent energy. This is contrary to the idea that photons are force carriers, thus implying that they are energy carriers. We shall say energy carriers, because it is more accurate, as will be explained momentarily. The author believes Einstein and Planck were accurate in saying that photons were energy. This is the truth that resides in nature. The proof is that bent energy has mass, and photons have mass. There is only one way to have mass, and that is to be energy, ergo photons are energy. The other thing we must come to accommodate is the fact that photons at rest have no mass. This implies that there is no bend in energy. Think about this for a moment. How do we know inherently, that photons are not really just force carriers? Simply put, light doesn't accelerate from 0 meters per second to "c", rather photons travel at c from the beginning of their existence. In short, photons do not accelerate. They instantaneously assume the speed "c". That begins by some subatomic activity. The photon has no mass at rest implies that the energy was bent in motion (when it had mass), and now it is not being bent while it is at rest. This gives rise to the idea that the energy of a photon is not permanently bound where the bound particles (i.e. the smallest part of massive hadrons, whether quarks or what not) are. What is causing the energy (that is the photon) to bend? What causes any energy in motion to bend? Once more the answer is perpenular friction. Energy in the universe is always spinning. This is the inherent nature of energy. All we need is something to bend it to produce a gravitational field. Energy within a relatively moving system falls subject to perpenular friction just as much as a particle with mass. In retrospect, what we have is energy moving through space at a constant velocity relative to everything at once and is susceptible to perpenular friction. The string of energy bends, and produces a gravitational field while in motion. To review, remember that the force of gravity is produced by the opposing rotations at the opposite ends of the string. We get concurrence via the second postulate of the Law of Fundamentals.

I would like to take this time to go a little further into relative motion and the relation it has to space-time. As one object travels relative to another, the object observed to be in motion causes the universe to vibrate. As the object moves faster the universe oscillates faster. This appears to have something to do with diffraction of light and the DeBroglie wave function, but as of right now the author is unable to understand it entirely. There is a vibration of space-time. The thing to realize is that if there is a maximum velocity ("c") then there must be an absolute maximum oscillation frequency of the universe. So the frequency of the space-time associated with the speed of light is the maximum possible frequency attainable by any object. As we will find, even this particular frequency of space-time oscillation is only achievable by electromagnetic radiation. Objects that assume some non-relativistic mass cannot achieve this particular frequency.

Up to this point, perpenular friction has not been explained in extensive detail. We will get to it, but we must first observe, hypothesize, and then prove either by induction, or deduction. This is the basic scientific method. Moreover, physics is an experimental science so you will have to make do with observations until a later time. Perpenular friction is a very elusive concept needing more tools to unlock the divine mystery. Perpenular friction will be more completely introduced in Chapter Five.

What we have stumbled across is that the makings of space, time, matter and energy are all the same thing. They are regulated by the same rules, and intrinsically even though they are not similar in their emergent properties. Much the same way all animals are made of cells and yet may appear nothing like one another, so too, space, time, and energy are the same thing, but in different forms. But there brews a sort of confusion on the quantum levels. All massive objects in the universe are known to be derived from what we call particles. Particles are manifestations of their own differentiated existence. The problem comes that these particles are nothing more than bound oscillating strings that make up everything we know to "exist." Simply said, there is a problem with scientific thought these days. These include the Holistic and Reductionist views. This the same problem Einstein, and so many others, had with the quantum theory, and the Copenhagen Interpretation. Even though Einstein was one of the fathers of quantum theory, and yet he refuted the quantum theory endlessly. One of the greatest stories in physics is the one of Einstein and Bohr at a conference. When talks of the quantum theory began he stated, "God does not play dice with the universe," which is arguably Einstein's most famous, or infamous quote. On the third time Einstein said this, Bohr stood up and said, "Albert, quit telling God what to do." This demonstrates the extent of the confusion when it comes to understanding such abstract concepts of our universe. Holistic, and Reductionist views begin to fail.

The moral of the story is not only did it take an astounding team of physicists to break the mold, of pre-quantum thought but also the unnerving revolutionary thought processing of Bohr, Heisenberg, Dirac and DeBroglie. It is the opinion of the author that these men broke a mold, which should have taken another hundred years, thousand physicists and countless experiments, to figure out. This is what we need in this day and age. The beginning of a new revolutionary track of thinking is the remedy. The walls between reductionists and Holisticicists must come down, because they are indubitably two halves of the whole.

It is of the utmost dire importance to remember that anything with mass is nothing more than bent energy. Keep this in mind when we discuss everything from quarks to bosons. Now that you have seen a small portion of how we will use the oddities in

Chapter two, it may be a good time to go back and look over the chapter. The definition of force was the key to explaining gravity. Many other explanations will come in very much the same manner. In all of this, the question arises as to how energy came to be matter, and what the relations between energy, and matter are? With that, we move on to the next chapter.

Chapter 4

Materialistic Separation

There are a few very awe inspiring concepts in physics that are currently unexplained. One of them was discovered by Dutch physicist M.J. Sparnaay called Zero Point Energy. The observance was this: at approaching zero Kelvin, a container of liquid hydrogen was being pumped out of energy. In theory, all motion should stop at zero degrees Kelvin. However, this was not the case. Sparnaay observed that the atoms actually began undulating. Amazingly, no known source of energy was to cause. This zero point field is supposedly everywhere, and capable of being tapped into. This implies a free energy source. As for the Law of Conservation of Energy, and Second Law of Thermodynamics, not much is said in terms of whether additions must be made to the laws, or if there are cases of exception. The major question is the source of energy. Some physicists speculate that there is a sea of immense energy just "floating" around, and some physicists propose that there is enough energy in ten cubic centimeters to fuel the United States for weeks. John Wheeler calculated that there was enough energy in one cup to evaporate all the oceans of the world. Research is vigorous to find this "free" energy, and as well as a way to make it readily available, if at all possible and assuming it exists.

The second unexplained occurrence dealt with black holes as it is unofficially accepted that black holes exists. Black holes supposedly lose mass over periods of time. Stephen Hawking of Cambridge showed that black holes "evaporate." Our interest is in how they evaporate. The following is a derivation of a black hole's escape velocity:

a. $v=\sqrt{(2mG/r)}$

As the mass of a body increases, so does the velocity, whereas the velocity and mass are directly proportional. The maximum value for the velocity is "c."

b. $c<\sqrt{(2mG/r)}$ As mass goes to infinity.

The problem came when math and indirect observation suggested that black holes lose mass. How could this be so if the escape velocity of black holes is "c" or greater? Hawking said that black holes emit this "Hawking Radiation." The problem is that the particles that are radiating cannot be from within the black hole, because nothing can escape the intense gravity of a black hole. Hawking's solution was that at the event horizon, matter and antimatter particles are being created. The matter particle is usually shot into space, while the antimatter falls into the black hole, and is annihilated with the matter in the black hole. This is a wonderful solution to a seemingly unsolvable problem. This leaves us with a net flow of matter out from the black hole, and the solution, which does not break any laws of physics. The question is where do these particles come from? Hawking and some others suggest that they are always there as virtual particles. This explanation does not appear to be completely correct. Particles that only exist in various instances, but are always there, seems to be a weak argument. The author feels that the idea of virtual particles is most absurd. The definition of virtual is "almost, but not really." Are there supposed to be all of these "almost but not really" particles floating around? This explanation is so poor that, as Wolfgang Pauli would say, "Isn't even wrong!"

These "virtual particles" tend to pop out of nowhere in other cases such as when there is a near perfect vacuum in space. Note that most of space is not a very good vacuum, due to an exorbitant number of neutrinos pervading space. The other observance the author would like the reader to take notice of is the occurrence of a photon's splitting into two particles (one matter and the other anti-matter). Afterward, annihilation of the particles usually takes place. It should be emphasized that there is an innate congruency between the observations of a photon splitting, and pair production of matter, and anti-matter. The key concept involved in these situations is energy.

All of this is occurring in a near perfect vacuum. The question is once more asked, "Where is the energy coming from?" The answer is space. The author would like to further this by saying that the composition of space, in its entirety, is energy. We have to look into the depths of nature and probe to find what space is. The answer is "nothing." This concept of "nothing" tends to stagger the human mind. The truth that one must come to is that the question in asking what space is can be best answered by saying space is not. The theoretical physicist's answer to our question is the pretense of existence. If we revert to the previous chapter, we have proposed that energy is nothing more than a one-dimensional string. Working backwards to the axiom that tells us distance is quantized, and we can extrapolate myriads of information. The most important concept here is that a three-dimensional plane can be broken down as an infinite number of two-dimensional sheets, and then breaking each individual sheet results in an infinite number of one-dimensional strings. Thus, we arrive at a more than shocking implication. All space is composed of one-dimensional strings, and when we add the idea of planck length we come to the truth that space is energy._This means that space is nothing more than an ocean of energy. A quantized cubic volume of space would in theory, contain enough energy to fuel any endeavor than one could possibly conceive. This is, because a cubic quantized volume would contain, and amount of energy approaching infinity. This is because a square that is "q" by "q" (lengthwise) contains an approaching infinite number of strings, where the width of a string is 0 (q/0=∞). Remember that infinity is a concept, and not a number. The calculations carried out by various physicists suggest that there is a finite amount of energy in a finite volume. The above theory suggests that in any finite volume, there is an infinite amount of energy.

We are not done. The origin of energy has been well explored, but why particles appear out of nowhere is not yet known. What we do know from Special Relativity is that matter is energy. We must again use one of our previously devised mechanisms in order to come to a solution. If we look to the Universal Principle of Waves, we see that in assertive wave behaviors, there are two extremes. Even though there is no medium, the second postulate of the Law of Fundamentals suggests that the system will attempt to reach equilibrium. The two extremes are existence and nonexistence. There exists instability in empty space, which increases as empty space collects, and gives rise to the emergence of energy. The amount of energy is determined by the instability of the region, which is partially empty. The larger the region of space is that is partially empty, the greater the instability. Moreover, an absolutely, or approaching absolutely empty region is severely unstable. In fact, the region is so unstable that the energy is completely extricated from space. The result is that the strings of energy are split into two pieces each (whereas the strings have zero width, and break no laws), and are looped into separate particles, matter, and anti-matter. The strings of matter and anti-matter rotate in the same fashion (there is only one way for string to rotate regardless of orientation), and are subject to the same gravitational force. Theories regarding antigravity are, in the author's opinion, a senseless waste of time. However, all possibilities must be exposed. Any observations of forces that oppose gravity are incompatibilities in wave functions, such as electromagnetism, or what have you.

Slightly unstable systems, such as the one seen in zero point energy experiments, are a little different. It appears that the significant lack of energy (in matter, and non-matter form) affects the system crucially. We will explore the reason for this momentarily in dealing with the Big Bang. The severe lack of anything and everything in a region causes the region of absolute empty space to perform a process I call Materialistic

Separation. This is when the instability of the universe is so great that energy is stripped from space proportionally to the instability of the region. The energy that makes up the universe as space is untapped energy. The author is sure that there is a particular satisfactory amount of space-energy, which is converted to a more conventional form of energy when the state of absolute instability is in effect. The energy that is taken out of space is tapped energy. We will use the terms, tapped energy, and untapped energy, to better differentiate between the physically occurring forms of energy, and from the energy that makes up space itself. Untapped energy will be the term used to describe space-time fabric, and tapped energy will be all other forms of energy, which includes matter. This brings us to an interesting topic, and the most predominant theory in astrophysics, the Big Bang. But first allow me to reassure you that I have not overlooked the classical laws of physics.

A great question that arises is about the Law of Conservation of matter/energy. Again we come to a confusing topic. This is that energy is not scalar, but rather it is a vector. This is very congruent with thermodynamics. Obviously, energy is not a vector in three dimensions. If it were we would see energy buzzing around in a weird fashion, which is completely uncharacteristic of our universe. Energy as a vector exists as energy in at least the fourth and fifth dimensions collectively. What is actually happening is when energy is untapped (energy in the form of space), it is parallel in orientation to the perpenular sheets. We will assume the energy strings are pointing in the positive μ direction. When energy becomes tapped, the energy is immediately perpendicular to the perpenular sheets, pointing in the positive Δ direction (if in the form of matter). In this case, the energy is in its most ordered form as matter. If the tapped energy is not in the form of matter, then the orientation with respect to the perpenular sheets is less than 90° to the plane. The less the angle that the energy vector makes with the perpenular frames, the less ordered the energy is, thus the less useful the energy is. Eventually, the energy will realign with the perpenular frames.

It goes without saying that this nearly abolishes the conservation laws unless there is yet another loophole. However, I do not think there is a loop hole. I suggest that a change be made to complete the Law of Conservation of Energy/Matter. While a system may continuously gain energy, the overall average usefulness of the energy within the system continuously diminishes. In other words, the average overall orientation of strings, with respect to the fourth dimension, approaches zero at all times. That is the case only if the Second

Law of Thermodynamics is retained, otherwise further changes must be made. This also gives way to entropy. So, if there is initially have five joules of energy in a system and four joules worth of useful energy. Then an instant later thirty joules of energy are added the amount of useful energy is some value less than four joules. The Second Law of Thermodynamics would prevail once more. As well defined the universe is in terms of symmetry, the author is confident to say that under no circumstances would the universe come to end in an energy death. Oscillation will occur regardless, and I feel that materialistic separation is a fail safe in the mechanism of the universe. Moreover, the Boltzmann-Lorenz Law suggested earlier would be put to the ultimate test.

Our discussion of the Big Bang will begin with the previously stated facts. Firstly, when a region is of absolute emptiness, the instability is maximum per unit of multi-dimensional space. The path thus far has been paved. We need only now to go into further detail, and we will do this at a later time. There is some constant that should be present in the amount of energy released under optimal conditions, thus telling us how much energy was released at the beginning of the universe. The author is confident to say (reasons to be stated later) that the universe produced many times the amount of matter needed for a Big Crunch as it is called.

Two other observations should be taken note of. The first is the Casimir Effect. The Casimir Effect occurs between two metal plates with which there are interactions present in between the two. The force can be expressed mathematically by the following:

$$F=(\pi^2\hbar c/[240a^4])A$$

Take into account that "h-bar" is Planck's constant divided by two pi. It is said that this Casimir Effect is the result of well interactions, and can be tied with General Relativity quite simply. The other observation is the Lamb Effect, which occurs between atoms. These interactions between subatomic particles cause fluctuations in the vacuum between them. Without having any further explored the subject, I would go as far as to conjecture that ZPE, materialistic separation, the Lamb and Casimir Effect are the same occurrences incognito. If General Relativity fits into all of this as easily as it is supposed to, we may be on the brink of a revolution much needed in a fairly stagnant time in physics.

We began this chapter in a resounding uproar of energy/matter relations and ending with a Bang, which we will further explore later. The important question comes in the materialistic separation and ZPE as to whether, or not, there is such a "free lunch" in the universe. The author is optimistic, while the classic laws of physics suggest otherwise. Many classical laws, because of the suggested theories, may result in a revisions and additions to particular laws of physics. The major concept that the author wishes to emphasize is that the universe began empty, and then was full. At least for a Planck time there was at best a compromise in laws. The author's hope is not that the past physicists were wrong, but rather that they never looked to the source of all energy, and because of that, adjustments must be made accordingly. These questions should be cared for as we discuss the terms of the Big Bang, and black holes.

Chapter 5

The Six Dimensions in Deftonic Theory

We should polish off the description of the naturally occurring force known as gravity. One problem in physics is that we can define a physical law of nature in an equation, and yet lack complete understanding of that equation. This leaves one with an innate unease within, and can leave one asking why a particular physical phenomenon occur. It is common knowledge that Issac Newton said that all matter has gravity, and gravity draws everything together in the universe. After Newton, Nikola Tesla attempted to add to the theory of gravity, but his work went ignored for the most part. Einstein furthered the theory with his works. The author has concluded that no one knows what gravity is, or that which makes it go, just as was the case with time. The sheer elaborate structure of gravity makes it impossible to make mistakes in its explanation. What is meant by that is there are very many cases that an explanation of gravity must successfully traverse in order to be correct, without error, or addition, or caveats. Such cases include verification of General Relativity. These cases also show the connection between Special Relativity, and General Relativity. Moreover, the ultimate unification lies between gravity, and quantum mechanics. We shall begin where we left off, dimensions.

If we take our five dimensional paradigm, and bend it as we have been doing, what exactly do we get? For reasons we will later explore, the dimension curls itself up into a sphere (the reasoning much involves the Big Bang). Since this is the case, we must revert back to our mental imagery of three dimensional space, that we are so familiar with. The continuity of space, as we know it, is some what distorted. This is, because the universe must now be composed of quantized packets of space. Furthermore, we must incorporate time as well. Therefore, the universe consists of quantized packets of space-time. The mosaic we must now carry in our mind is a universe of particles, which the author refers to as "bubbles." Hence, we have exceptional Euclidean problems. Picture this. You must take a sheet of paper, and cut circles in it. These circles will be where space-time exists. There persists a problem in the continuity of the space-time. There are cracks in the region where space does not exist. These little abnormalities are pieces of absolute and perpetually non-existing points, which will be called "constituents of the void." This is where things become very weird. We must expose ourselves into an entirely different way of thinking. Outside of our system which we call the universe, we can make no assumptions, because we have no knowledge that extends beyond our universe. Thus, the farthest we shall venture is to this boundary.

To clean things up a bit, I will define the "constituents of the void" to be a subset of the universe that lacks the parameters for existence. Those parameters for existence are dimensional space. I will go further by defining the void as being that which has the universe as a subset. Not only is the usage of theoretical mathematics correct, but it also suffices in differentiating the three entities, existence, the void, and the "constituents of the void." Also, the void, and constituents of the void are unquantifiable. The universe is observably uncountably infinite.

However, as far as our experience tells us, the universe appears to be as consistent as any Euclidean plane. Two things must be kept in mind. The cracks (constituents of the void) deal in the quantum world, and they are points where space does not exist, and renders our senses helpless. Basically, the universe appears to be as continuous as our experience would wish it to be. This brings us to the luminiferous ether. Michaelson, and Morely used an interferometer to show that whether the luminiferous ether existed. So are the previous thoughts on space-time completely null, and void? Has the author made a mistake? No, not necessarily. As a matter of fact, the experiment only showed the relations between space-time and light's travel. Einstein once said, "I did not say that the luminiferous ether did not exist, I merely said that it was not necessary in light's travel."

It is the writer's belief that the physicists in the early 1900's gave an improper description of the ether. By giving an incorrect description of what the luminiferous ether is, the profile of attributes that the luminiferous ether would take on, were meaningless. It is analogous to giving a poor description of a fugitive. Although the person fitting the profile was caught, the real fugitive remains at large. Nikola Tesla's theory of gravity, produced in the late 1800's, also included the ether. Nikola Tesla made it clear that the ether he was speaking of was not a rigid body such as a gas, even though he could not give a very lucid description.

First, we shall paint a mosaic of our universe's composition. We have the original five dimensions, and the adding of a sixth dimension, which gives us curled up sphere of space-time, or packets of space-time. These bubbles (as the author calls them) are planck length in diameter, and they are jammed into regions adjacent to one another.

Thinking back to the strings, remember that they have a rotation such that when they are bent, the opposing ends have opposing rotations. The Law of Fundamentals by which our universe is governed, appends the extra added twist that the string will actually want to cancel out, but cannot due to planck distance (a.k.a. planck length) restriction upon space. This gives rise to separately existing bent strings, and the fact that they will have an attraction for each other. All strings will provide an equal, and opposite for all other strings when bent. The key is the rotation of the strings. Because there is an instability, such that there is a compatibility in the wave functions for one another, the universe in this case presents means by which equilibrium may be reached. As the strings rotate, the adjacent bubbles rotate. Therefore, all of the adjacent bubbles that are rotating cause all of the outer bubbles in the region to rotate. With this rotation of bubbles comes a force upon matter equivalent to the radial velocity of the bubbles. Moreover, the rotation of one bubble will be rotating in an opposite direction with respect to the one radially behind it. This means bubbles will condense to the next lower radius of the circle. The result being that the existence of one bent string causes all existing bubbles within the system to rotate. The fact that these bubbles are changing position with respect to masses is what causes accelerations of mass, otherwise defined as a force. By this we see that the gravitational field extends to the entire universe.

The reason for the dampening of the gravitational fields, is because there are more bubbles to rotate further out from the center of mass than toward the center. For particular reason's we will not touch on those subjects. It is only important to know how the gravitational field works, and how all of the ideas proposed in these volumes mingle into a single theory.

A concept which has been postulated in this book is perpenular friction. We will now uncover the mystery as to what it is, and use it to tie the two Theories of Relativity together. As you know by now, the dimensions of space, sixth and above (if there are more), are wound into a sphere. Particles that exist are restricted to existing in bubbles, thus being a fragment of our space continuum. For intents and purposes, assume that the particles we are dealing with be a little larger than the bubbles, and neglect the mass of the particles temporarily. In relative motion we will travel from bubble to bubble. The definition of travel is, to exist at every bubble at some time along the path between two points. However, the particle will be found in the oddest of places along the course of its trip from "a" to "b" (should the particle ever reach "b"). As Richard Feynman had suggested, the particle takes all possible paths. This implies the classical definition of "travel" loses some, if not all of its meaning. There is an inherent reason for this, and that is that all points in space are the same point. Feynman suggested that a particle will take all possible paths, when traveling from "a" to "b". This may, or may not be difficult to understand. However, a means of better theoretical understanding has been constructed, and it is called Regressive Geometry. Only a few of the concepts of Regressive Geometry will be deployed. The use of Regressive Geometry is useful in gaining a better understanding of higher dimensional space theory.

This book shall only state three points in regressive geometry. The first being regressive time. Regressive Time deals in differentiation of events that occur in the same frame without change. In other words, when time does not exist, and events are jumbled together, regressive time separates the events for us. Logic is involved, and the importance of this type of time is to help make sense of such happenings as the Big Bang before time began. When some type of chronological notation is given in these cases, beware of regressive time. In such cases all of the events happened at the same point, not in time. Regressive Geometry is just a way to explain the "chronology" of events in theory. One must realized there is no such concept as chronology when time does not exist. Therefore, Regressive Time is to be used to build a picture. Another difference between regressive and Euclidean geometry is that the shortest distance between two points is no distance. In that respect regressive geometry is the same as Riemann geometry. However, in our new geometry we go further, and say that all points in space are the same singular point in space.

The problem with this theory is that the amount of useful energy must have been zero before the Big Bang. Would this then imply negative amounts of energy? By no means is this the case. One must first ask how sensible of a question is asking about the entropy of the universe before the Big Bang. In essence, it is the same asking questions about the universe before it existed. There must at least be one exception to the Second Law of Thermodynamics, or else the Big Bang would not have been able to occur. Perhaps there is hope for those physicists working to make perpetual motion machines. There is so much work that needs to be done in this field that not much more can be said.

A multitude of different experiments have illustrated that the quantum scale of physics operates under a different set of laws than the macro universe. The physics of large scale, and quantum scale appear to be very different. In Thomas Young's wave experiment, one needs only to shine a source of light at two slits, and the result is diffracted waves. Remember that light is also a particle. Is a single particle hitting multiple locations at once? No it is not. It is one spot being hit by one particle. One must unify the locations. The problem here is that physicists had two choices, and took the more obvious one without considering the other. Instead of unify points in space, they said that single particles split, or duplicated, and strike multiple locations. Taking a look back at Chapter Two, the rule for wave mechanics says that in assertive wave behavior there is no medium. Moreover, the satisfaction of the dynamics can only be obtained by unifying the extremes. In this case, the extremes are positions defined by energy.

An example is that of the photon. Some time ago, there was an experiment dealing with splitting photons. Everything that one photon did, the other would mimic. What has to be realized is that these reactions between photons are occurring instantaneously. If someone were to apply an electron volt to one, the other experiences the force immediately. By saying that these are two differentiated particles, then surely the instantaneous reactions would not be experienced so quickly or even at all. At most the transfer of energy would occur at the speed of light. Now if we were to look at the two sets of positions in space as the same, then the results in this experiment become clear. The Deftonic Theory suggests the that reason the particles then experience the same forces instantaneously is, because they are the same particle.

If we send some particle through space from point "a" to "b" in a straight line, we are disturbed to see that the particle is not where we expected. When I say expected I mean by standards of Newtonian motion, or relativistic motion for that matter. The first thing is that nature gets carried away in accounting for energies. As long as the total energy is accounted for at quantum levels, it does not matter the form that the energy is in, so long as the First Law of Thermodynamics upheld. Then we have Heisenberg Uncertainty, saying that we can never know the energy of a system entirely. Basically, we must now name our particle traveling from "a" to "b", Jim. Jim has some energy, both potential, and kinetic. The problem is that a second particle named Bob exists. Bob has some energy, both kinetic and potential. As long as all the energy is accounted for in the system, it is absolutely irrelevant what happens. So in this case, it does not matter what the individual energy of the particles is. Only the total matters. The particles could very well trade places or the Bob particle could lose his potential, and Jim could gain kinetic energy. Because of this sort of interconnectedness, many of our macrologically existing laws are compromised.

The Heisenberg Uncertainty Principle, comes from the following scenario. The equation for circumference of a circle is:

1. $2\pi r=C$
2. $\pi q=C$

We substituted (q/2) for "r," because the diameter of each bubble is q and half of that is the radius. An inherent problem persists in our scenario. It is alright for some of the "in-between mathematics" to lose some meaning, but our end result must be consistent. We see the circumference equals πq, but as we very well know, an irrational number times a rational number equals an irrational number. In layman's terms, if we have a bunch of strings "q" long, we will never be able to physically construct the circumference of these bubbles, which are perfect circles. Heisenberg uncertainty merely hides the absolute measurability of the energy. Nature is actually taking into account the fact that the true measurability cannot be made absolutely. We will always get an excess of string or an incomplete circumference because π is irrational. We will always be about .14 units of "q" short, or about .85 units of "q" too long. What some people don't realize is that position defines potential energy and that is the other half of this story. Kinetic is only one part of the energy game. Not knowing the absolute energy means that the position is unknowable in full accuracy. Moreover, if absolute energy is unknowable, then the momentum of a particle can only be measured with a certain degree of accuracy. The accuracy at which we can measure the change in position, and momentum is: $h/(2\pi)$, where "h" is planck's constant.

$$\Delta x \Delta p \geq h/(2\pi)$$

The reductionist's view of chaos may begin here for those reasons.

Getting back to the particles hopping around in space, we come to quantum tunneling. The jumping around of particles, involves the same principle that every point in space is actually one in the same. Many particle physicists have observed electrons go through barriers, and they appear to be breaking light speed. How can this happen? It is known that motion consists of traveling over some distance in a particular amount of time, and existing in every position in some quantity of time. With the paradigm that we currently have painted, we must also remember that there is a macrological and micrological sense of time. Time is completely contained within each individual bubble, and yet is a macroscopically observable entity. As a particle changes its position from bubble to bubble, a particular number of macrological perpenular frames are passed. The perpenular frame by the way, is the universe encompassing slab that accounts for all the bubbles. It bends at the will of each individual time plane. The "planes of time" is just the idea that object A has passed through more bubbles than object B. A therefore is older than B (disregard dialation). The problem is that electrons pass exorbitant amount of space in a miniscule amount of time. There are many implications, and conclusions that can be arrived from this theory of multidimensional space. However, Perpenular Friction has not been fully explained. Therefore, this multidimensional theory cannot be fully explained until a final picture Perpenular Friction in its entirety has been completed.

It is known that space-time resists the difference of inertial reference frames, otherwise known as perpenular friction. It is easy for an individual particle to go anywhere, because the potential energy is very loosely defined. The key factor is that a closely regulated potential for energy is open to very little change. An example is in an atom moving through space. The electrons go anywhere (for the most part) and everywhere. The neutrons are a completely different story. Very little or no positional changes occur relative to other nucleic parts. This is because the structural continuity would otherwise be compromised. The fail safe for the structural integrity in this case is the "strong nuclear force." The result being, that the particles cannot follow the path that they would most desire to follow had they not been bound to the atom. The strong force sets up a region where large amounts of energy would be needed to compensate for the loss of even one particle. The particles pull, and drag along space and time, because little shifting occurs in the relative positions of particles in the nucleus. The other thing is that the bubbles are "stretchy." They have an innate volume, but that is easily compromised with the addition of a little energy. This is also the other reason that there is some tug on the universal fabric. It takes some energy to stretch the next bubble for an electron to occupy that particular position. This is the realm of momentum. A seemingly classical concept leaps into the Einsteinnian world, and beyond. We can view perpenular friction as being a bulk, or excess of energy in a region, because that truly is what is happening. Any relative motion gives the energy bulkage needed for such observations in time. As we very well know, these bulkages of energy are more apparent at higher velocities. The bulkage curves the many dimensions of space, and the curve makes the kinetic energy in the miniature system bend. The miniature system being the one that is moving. This in turn, causes the strings that account for kinetic energy to have some gravitational field. Along with that, the increase in mass regulates the speed of the object, because objects with a large moment of inertia are difficult to accelerate. With an appreciation of physics, one will immediately see a parallel between our current situation and capacitors. When a capacitor posses a certain critical amount of charge with respect to a dielectric's strength, a dielectric break down occurs. The region can no longer capacitate the energy bulkage. This simply means that the charge will flow over the dielectric. Very similar to that, the energy within a region builds to a critical point, and we have a dissipation of energy.

This is exactly what happens to our previously mentioned electron in tunneling and tunneling is the universe's system of energy dissipation. To say that the electron is traveling through the barrier is incorrect. The definition of travel inherently, means to pass through every adjacent point between two points. What we have in tunneling is a dissipation of energy, thus cutting out some of the distance. To demonstrate this, assume there is a particle, Mary (assumed to be larger in volume that planck volume) and she is initially at point "A". She wants to go from point "A" to "Z," but there is a wall from point "L" to "P." If Mary is accelerated to nearly the speed of light, at some point before "L" Mary will dissipate some energy to appear in some point between "Q" and "Z". Mary stretched a single bubble at planck volume to a volume suitable for containing her, and transported herself over some distance relative to the dissipation of her energy. The following is the suggested relation between the energy Mary dissipated and her distance traveled;

A. E/q_e=# of strings

Where E is energy dissipated and q_e is the smallest portion of energy.

B. (# of strings)$q_d=L_o$

Where l_o is the original distance equivalent to energy by proportion. Lorentz's equation must be added. We start with the following;

C. $L\sqrt{[1-(v^2/c^2)]}=L_o$

D. $L_0/\sqrt{[1-(v^2/c^2)]}=L$

E. $([E/q_e]q_d)/\sqrt{(1-(v^2/c^2))}=L$

This is only a theoretical relationship, because experimental confirmation has not yet occurred. L will be the observed distance after the energy is dissipated in an Einstein-Rosen bridge, or in quantum tunneling (the same thing). It turns out that this mode of transportation was actually arrived at already.

They called it an Einstein-Rosen bridge. Of course these require a much greater abundance of energy, because they are the macrological equivalent. What would happen to an object like a large car if we sent it approaching the speed of light? If we had an infinite amount of energy, would the car just go faster, and faster through space? One must agree that at some point the density of the car will reach the critical density of a black hole. Especially because the volume will decrease with speed, while the mass increases with speed. This is a wormhole, or Einstein-Rosen bridge. It is made possible by the fact that all positions in space are actually one in the same. As soon as the volume of the object is equivalent to the volume of the bubble of space, transportation occurs. The added energy will increase the volume of the bubble, as the length of the object contracts as the velocity increases. The object will dissipate it's energy over some distance when the two volumes (object's, and the bubble's) are equivalent. This will most likely occur a critical density, as demonstrated by black holes.

The Higgs field, and Higgs boson theory are nothing more than what the author has already stated. The Higgs boson is nothing more than packets of quantized space-time. They are infinite amounts of untapped energy, which compose the fundamental parameters for harboring tapped energy. The Higgs particle that particle physicists are looking for today is actually the graviton. In short, the author of this book believes the graviton that particle physicists are looking for does not exist. So anticipation of the Higgs particle discovery may not be wise, because the experimental physicists are looking in the wrong direction. Moreover, it does not appear that any physicists have realized that the graviton, and the Higgs Boson are the same particle. The Higgs field is the collection of all of the "Higgs Bosons," or as the author calls them, bubbles. They are differentiated by the energy potentials at different regions in space. The Higgs field is what gives an object its mass. The more interaction it has with the Higgs field, the more mass it has, and the converse is also true. The math is somewhat tedious, but the mathematics that the theorists have come up with will be displayed for those who are interested. We start with the U(1) theory, which deals in the spontaneous symmetry breaking which is called the Higgs mechanism.

A. $D_\mu=\partial_\mu+iqA_\mu$

If we add the Langranian of free gauge field, A_μ results in;

B. $£=D^\mu\varphi D_\mu\varphi-V(\varphi)-¼F_{\mu v}F^\mu v$

The Langranian is now invariant;

C. $\varphi(x)\rightarrow\varphi'(x)=\varphi(x)e^{iq\varepsilon(x)}$

D.$A_{\mu}(x) \rightarrow A'(x) = A_{\mu}(x) + \partial_{\mu}\varepsilon(x)$

The Langranian in terms of variables σ and η;

E.$£ = \frac{1}{2}\partial^{\mu}\sigma\partial_{\mu}\sigma - \lambda v^2\sigma^2 + \frac{1}{2}\partial^{\mu}\eta\partial_{\mu}\eta$

Somewhat obvious is the fact that an extra degree of freedom can be absorbed.

F.$\varphi(x) = (\sqrt{(2)}/2)[v + \sigma(x)]$

The Langranian is left as;

G.$£ = \frac{1}{2}\partial^{\mu}\sigma\partial_{\mu}\sigma - \lambda v^2\sigma^2$

We can imagine (using regressive geometry) the universe as a microscopic whole. Approaching an infinite number of bubbles across what we image as being the macro universe, a seemingly smooth plane. However, each bubble is the same bubble, even though they are in different positions in space. If we imagine a sheet of bubbles going from point "A" to "B," we could add imaginary tunnels connecting every individual pair of bubbles, thus showing that one bubble leads to all other bubbles. Between the stretching of the bubbles, and the decreasing relativistic volume of objects, it is fairly certain that this Einstein-Rosen bridge will occur. The other thing that needs to be elaborated on is tunneling. Einstein-Rosen bridges, and tunneling are the same thing. The only difference is the amount of energy involved. Of course it is very easy to see that the less volume a particle has, the easier it is to stuff it into a single bubble. Also, the more mass the object has, the more energy required to fit the object into a bubble. The suggestion that anything can truly travel faster than light speed is preposterous when one considers the Lorentz transformation. Lorentz's factor shows that light speeds are unachievable for bodies with some mass. When objects move through space relative to another object, the fabric of the universe actually oscillates. When something travels at light speed, space-time is oscillating at the greatest possible frequency. We could spend the rest of our lives describing the mechanics of motion in its entirety, but it is important that we know why the speed of light cannot be broken. The details are for experimental physicists to work out. When you add energy to beams of photons, they will certainly behave the same way. It should at first glance, be obvious that these particles are not traveling beyond the speed of light. The truth is this assumption that particles exceed the speed of light is a certain fallacy. The particles are skipping across space and continuously dissipating energy. Thus, we shall observe the photons having crossed (not traveled) a distance faster than light could have. This is all thanks to Einstein-Rosen bridges, otherwise known as wormholes.

The other important idea is that "Zero Point Fields" are nothing more than an empty Higgs Field, or a nearly empty Higgs Field. The Zero Point Field is just space-time in an unstable region in space where the prerequisites for ZPE, Materialistic Separation and all other forms of spontaneous energy production, are present. The universe is nothing more than a Zero Point Field that is for the most part stable, thus unifying what we define to be space time, and Higgs' Field.

The most intriguing part of quantum mechanics is the Einstein-Rosen-Podolsky paradox. We have actually referred to it already without mentioning it. We talked about the split photon, which is a case of the EPR paradox. In this paradox a particle is sent down a shaft. The shaft comes to a fork, and the particle must go down either one path, or the other. So which shaft does it take most often? Human sense suggests that it would take one, or the other. The reality is that the particle hit the targets in both shafts at the same time. However, the particle does not appear to be split, or altered in anyway. The particles are the same particle, as far as can be observed. How can this be? Is it not contradictory? For something to exist, does it not have to exist at one particular place at one particular time? Time's contradiction only poses a problem indirectly. Again, if we reach back to the very fundamental implication of the sixth dimension, we see that this is very much possible. Through further examination, we see that the uncertainty principle endorses it. The answer to this question is, when more than one possible outcome resides on the question of energy, all outcomes must occur, and all possible paths are taken.

In going further into the EPR paradox, we find that not only can a particle exist in different positions depending on energy, but in displacement problems there is another interesting occurrence. As Richard Feynman put it, when trying to figure out which path a particle will take in its displacement from "A" to "B", give up. A subatomic particle takes all possible paths, and this explains such unusual laboratory, and mathematical observances. The one of the most important equations in quantum mechanics is the Schrodinger Wave Equation (1926), which was adjusted by Paul Dirac (1928).

A. $Ih(\partial/dt)\psi(x,t)=(-h^{2/}2m)\Delta^2\psi(x,t)+v(x,t)\psi(x,t)$

(schrodinger's original equation)

This is mostly from energy, and momentum conservation and in assuming the energy is evenly distributed in space we get;

B. $\partial E/\partial t=0$

C. $\partial P/\partial t=0$

D. $VE=0$

In an isolated system of such we can go further using vectors;

E. V X P =0

The next step is not very obvious but it can be easily computed;

F. $(XP)/|X||P|=1 \Rightarrow de^I/dt=0$

G. V X $e^I=0$

Now we must take the derivative of ψ;

H. $\partial\psi/\partial t=(EI/h)\psi/\sqrt{}(1+[(Et-RP)/h]^2)$

Now we shall make one final assumption (I) and follow it with a series of variations.

I. Et-RP=0 (as do the vectors RP)

J. If $\partial Et-RP/\partial t=0$

K. $E=(\partial R/\partial t)P$ or $(\partial R/\partial t)=E/P$

L. $\partial\psi/\partial t=(EI/I)\psi$

M. $-Ih(\partial\psi/\partial t)=P\psi$ or $P=Ih\Delta$

If we square this operator;

N. $P^2=(mv)^2=2m(mv^2/2)=2m(KE)=-h^2\Delta^2$

It is important to remind the reader that the Hamiltonian equals potential energy plus kinetic energy.

O. $H\psi=-Ih(\partial\psi/\partial t)=-h^2\Delta^2\psi+\Delta\psi$

Schrodinger's Wave Equation describes 0 spin angular momentum. We finish up with Dirac's spinor;

P. $ih(\partial\psi/\partial t)=(\alpha cP+\beta m_0c^2)\psi$

As you can see the positions in particles, and the certainty that they will be in a particular position, becomes little more than asking how many angels can dance on a pin head. In applying the "bubble theory" we see that not only is the particle in question traveling all paths possible, but there is only one thing that differentiates these various positions in space (bubbles). The one thing that separates the bubbles in space is distance, energy, and time. Other than that, all positions in space are the same quantized bubble of space.

A very intriguing thought experiment that supports these theories is in taking photons in white light, and splitting the photon into two. In this case, one is green, and the other is magenta. When the green "piece" is put under red glass and observed, the other particle becomes blue. One should notice that in all cases the photon's color, when added to the other particle, was white? They are the same particle, and this shows that space is one bubble duplicated into many. Still, it is only one bubble.

How do we use these bubbles to tie the theories of relativity together? Let us draw two pictures. The first is a ship traveling through space. The initial background is a flat uniform sheet of bubbles. As the ship passes through the bubbles, the ship drags space. The other thing to take note of is that time runs slower at the rear end of the ship than the front, where the ship is directed. Also, there is a blue shift toward the back of the ship. These theoretical "observations" were collected by Einstein in what he called the Equivalency Principle of General Relativity. If we take a large spherical mass that we consider to be stationary, we see many similarities. We take the same uniform background of bubbles. The observation is that time slows down as you get closer to the center of the spherical mass. One of two things is happening. One solution is the sphere is rapidly expanding, which we can disprove by looking at the gravitational field equation:

$$F=(Gm_1m_2)/r^2$$

After seeing this, it becomes very obvious that the most perpetual force in our lives would become infinitesimally small if such happened to the Earth. So if the spherical mass's circumference is not increasing, then the bubbles must be crossing the surface by other means. But we have agreed that bubbles must be crossing the surface of the sphere. The only other possibility turns out to be very simple. The bubbles must be moving toward the center, therefore they must themselves be moving to the center of mass. The intention is to go toward the center of mass of each individual particle, but the whole reason these bubbles are moving is due to bent strings. When the strings are bent, and spinning, the surrounding bubbles are moved by the spin, and sent in a direction that would put the bubbles in the center of the ring of string. The result of having multiple strings looped in the same space-time, and close to one another results in a center of mass location. This occurs because the bubbles attempt to unify at one loops center, but they continue on due to the attraction of another until the attractions are all balanced out in a particular location named the center of mass. It is at that point where all mass in the system will converge, and all bubbles.

What about time slowing down near the center of mass in our sphere? If we place the bubbles uniformly outside the sphere at radius "x," and they move to radius "z," which is closer to the center of mass, a problem exists. That problem being that there is no way to fit the number of bubbles at radius "x" into "z." Therefore, they have to crunch down, and objects closer to the center of mass will exist in less packets of space-time. What we have is at each individual radius is a differing number of bubbles at each shell existing. An object in the outer reaches of a gravitational field will observe a faster pace of time than the object at a lesser radius of the same gravitational field. This again is, because of the unification of these packets of space-time at the next lesser radius.

This ties in with observers in relative motion, and their observations in time dialation. We will think about a space shuttle traveling through space for the unification of Special and General Relativity. The last points on the space shuttle causing perpenular friction (dragging space and time) will be viewed as an imaginary tangent to an imaginary circle (assuming the space shuttle has a flat rear wall). Of course this has already been done by Einstein himself. The scenario we will next use will need the absence of gravitational fields to emphasize the behavior of the bubbles with respect to the space shuttle's motion. Assume now that object "A" has no mass, and is a point object. This is important, because we need a uniform space-time, and an object to be in relative motion with. What we have is a well packed region of bubbles uniformly distributed. Now we add the space shuttle moving at a relative speed away from the point object "A." The realization that must be made is that the space shuttle is passing through bubbles, the same radius differences are occurring due to the perpenular friction. The gravity well is changing magnitude in space as the shuttle accelerates. Because of this, the bubbles that the ship is passing through approach convergence. The slower the object's velocity, the larger the imaginary circle that sits on the tangent, until the object is relatively motionless, and the tangent becomes a flat line. If the object were to reach the speed of light, the convergence of the bubbles would be at the point where the tangent line meets the imaginary circle. Thus, we have the unification of the Theories of Relativity.

To summarize the two theories, the key was the idea of completely quantized packets of space. These quantized packets of space-time were implied by the sixth dimension to be spherical in nature, thus the name bubbles. While the details of the sixth dimension are not very well defined, it has given us more than enough information to solve many problems plaguing physics to this day. Many different numbers have been thrown around in attempts to figure out how many dimensions we live in. A six dimensional space has been one of them. The sixth dimension is very intriguing, because many problems we currently have today are answered in its exploration. It is the author's personal belief that there are only six dimensions, which compose the universe, and the reasoning will be given in the Seventh Chapter.

The simplest implication from General relativity is the gravitational lensing of space-time, and how light travels, because of it. If we look at Einstein's classic idea of space-time being a rubber sheet, the sheet bends when masses are added. The other thing is that light travels at a constant speed on the sheet, and is known by the Compton Effect to have a relativistic mass. Then the light would travel a curved path through space, around masses. This is most observable in light passing stars. Sir Arthur Eddington was the first to observe this during an eclipse, and in doing so proved Einstein's theory. All that needs to be done is to add the bubbles to this theory to give it the finishing touch. But first it should be pointed out that in a force field, the bubbles change positions "faster" where the field is stronger. When velocities are used dealing with bubbles, the bubbles change position in some amount of imaginary time. It would be meaningless to say that the bubbles moves without respect to time.

As a particle of light passes a massive object, the photon exists in bubbles. Therefore, it passes through some number of bubbles in passing the mass. The "catch" is that these bubbles are changing position with respect to the mass, and in fact moving toward the center of mass. We can imagine that in every bubble the photon passes through, the respective bubble carries the photon with it toward the mass. Thus, it is given a curved trajectory in passing the mass.

The most important idea we talked about was the quantized packets of space. These spherical bubbles are the smallest constituents of the macrochasm we call space-time. We also uncovered the reason for the EPR paradox, and Einstein-Rosen bridges/quantum tunnels/wormholes, which are the result of all the bubbles being one unified, and quantized packet of space. The Zero Point Field and Higgs' Field were shown to be the same thing. Moreover, the Higgs boson has been shown to be the graviton, which has been called a bubble.

If we look back to the origins of gravity, we see that the cause is bent strings that spin. Moreover, because of this spin the bubbles of space around them are caused to unify at the center of mass or converge. The sole reason such potentials of gravitational energy is set up is, because of compatibilities in wave functions When moving bubbles cross a beam of particles, let us say photons, the bubbles cause a change in trajectory of the particles. It is for this reason that gravitational lensing occurs. In this chapter we also unified the Theories of Special, and General Relativity. The unification of the Theories of Relativity explain similarities in time dialation, and the Equivalency Principle. Maybe the most important thought is that quantized strings cannot make a completely self consistent circle, or sphere. This gives rise to the Heisenberg Uncertainty Principle. With all of these theories, and thoughts tied together we need to look at a few other implications.

Chapter 6

Implications of Deftonic Theory and Proposed Theories

With any new scientific theory, it is always good to pluck out a few implications to see where the theory is going. So in this chapter we will go over some of the more obvious extrapolations of the theories. We will jump around a bit more to familiarize the reader even more with the previously covered material.

The most profound part of the theory (besides unification of Relativity) is that of energy. The thing that must not get past the reader, is that energy is quintessential. It is not certain where the mathematicians have gone wrong, or at least have incomplete calculations, but an error has been made in energy calculations. The physicists seem to believe that in the volume of a cup containing a vacuum, there is enough energy to fuel the world for months with the concept Zero Point Energy. The problem with this is as follows; in a quantized packet of space, there are an approaching infinite number of diameters through it. Each one of these diameters is of planck length, which is the smallest portion of energy possible. Therefore, in one bubble there is an approaching infinite amount of energy. The other thing is the contradiction astrophysicists make in description of the beginning of the universe. They say the universe began infinitely dense as a "point." The biggest thing is that most physicists working with Zero Point Energy is that they have not yet realized their source of energy is the fabric of the universe itself. The other misconception many physicists have is that these Zero Point Fields are not ubiquitous. Moreover, many physicists believe Zero Point Fields are dense regions of usable energy, which are floating around. According to the Deftonic Theory, this is also incorrect.

Yet another thing about energy is that it is a one-dimensional string. That is difficult enough to accept. But what is a one-dimensional string? It is only observable in the remnants of its interactions of energy alone. The strings of energy in essence can be traced back to the point from which they are derived. The unrelenting truth is that energy is nonexistent. If we look back to Chapter Two, we see that all that exists fundamentally has an equal, and opposite that have a natural propensity to reach equilibrium. Is there something wrong with the Law of Fundamentals? Absolutely not! This is the very reason we have symmetries prevailing in every portion of physics. All that exists in the universe, by definition, is derived from energy, and actually is a form of energy. Since energy has no equal and opposite, it therefore, does not exist. Again, let the reader be reminded that dimensions are merely prerequisite for existence, and energy is merely a side effect of these particular happenstances. The universe's nature is to counter the nonexistence of the universe, so the parameters must be set to contain such an idea as existence. Energy truly is a remarkable aspect of the universe. We will go into further detail in the next chapter.

If the reader has not noticed by now, the entire cause of time dialation is due to the stretching of distance between perpenular frames. As an object accelerates, it must still pass through every bubble between two points. The drag on space-time is due to the paths that the particles (that compose the object) would take in nature, were they not bound, and the restricted paths through space that they take. Because the object has accelerated, the difference in inertial reference frames causes a buckling of space-time like the rubber sheet analogy given by Einstein. The particles are bound together by forces, so they give little change in position. Hence, there is a somewhat stable structure. The particles have a natural desire to travel in all possible paths to that point which is restricted. Since the most desirable path, or the path of least resistance, cannot be followed, bubbles take the path of resistance in which bubbles stretch. The final thing is that in these stretches of space we also stretch time, because space and time are intertwined in the sixth dimension as one whole known as a bubble. One cannot be stretched without the other. This also accounts for similarities in the equations. The result in perpenular frames is the elongation of perpenular frames. The simplest example is of light. The perpenular frame of a photon is perpendicular to those of two objects that are relatively at rest with respect to each other. While the distance between frames remains constant (planck length), the perpenular sheet increases its area and the angle of the fourth dimension with respect to the fifth dimension decreases. In relative motion, the distance between perpenular frames change.

Regressive Geometry, as the author calls it, begins to take shape in this theory to describe the universe. To date, it has not been completely developed, but a few key theorems protrude. One key factor that must be realized is one of Richard Feynman's thoughts. He said that a subatomic particle going from point "A" to point "B" takes all possible paths to get from A to B. This has been observed monotonously in quantum mechanics. The explanation of this is a two-part answer. Firstly, the shortest distance between any two points is zero units of distance. Boiling over and into the second theorem is that all bubbles in space are the same bubble.

With Regressive Geometry, we see that the only difference in two positions in space is potential, and kinetic energy. Due to the fact that all the bubbles are the same bubble differentiated by distance, we have built an imaginary web of tunnels. This is the beginning of the wonderful world of quantum mechanics, which is the most difficult to understand. From each bubble we can say that there is an imaginary tunnel going from it to every other bubble in the universe, much like a map of all highways. We explain this using the quantum tunneling, which is nothing more than an Einstein-Rosen bridge. When these bubbles are stretched to engulf an entire object, the object will not need to exist in every bubble between two hypothetical points, "A" and "B." This is also why it appears that particles travel faster than light. The problem is the definition of "travel." Travel is done when an object exists in all locations between "A" and "B" at some point in time. Travel is not what occurs when a particle passes through an Einstein-Rosen bridges, or a quantum tunnel. If the particle tunnels, then by definition, it has not traveled the distance. It has skipped over this distance. This gives much hope for future space travel, but at the cost of much energy, depending upon the distance tunneled.

There may not be any more than six dimensions. The entire purpose of the dimensions is to contain an existence fully capable of oscillating in every respect. The thing that eludes reason is that there are string theorists who have models containing thirty-two dimensions, or more. One of the leading string theories of this time contains thirty-two dimensions. In fact, the theory has been included into the M-Theory, which is the all-encompassing string theory at the moment. It is senseless attacking such problems using blind mathematics alone. Mathematics can be virtually meaningless when a physicists derives mathematical expressions, which describe higher dimensional spaces. This is not to say that there are not 32 dimensions. Since math is so difficult to translate into meaningful physics, it is difficult to infer all of the importance of the sixth dimension as it pertains to the universe. If any progress is to be made, we must devise a system of understanding as to how the higher dimensions are constructed. The flatlander theory is an antique, and impossible to use for the further exploration of thirty-two dimensions. Mathematics is a perfect science to describe an imperfect world. However, math alone cannot be trusted. Mathematics describes not only this world, but also all possible universes set by similar rules. It is a matter of differentiating mathematical lies from mathematical truths, as they pertain to our universe.

To reiterate the Einstein-Rosen bridge idea, assume some object with a mass accelerates toward the speed of light. It is very well known that the object will never reach such a speed. As it approaches the speed of light, the volume decreases via length contraction as the mass increases. The result is an extraordinary increase in density. What happens when the object reaches the density of a black hole? The universe has an energy dissipation system known as Einstein-Rosen bridges, which is the same as a small scale version called quantum tunneling. Of course much less energy is needed to observe an electron tunnel through a barrier, but this occurs at macrolevels as well. The understanding that the universe is composed of packets of space is crucial. All objects exist in bubbles of space-time. The particular bubbles in which an object exists changes. But that is where the micrological understanding of time comes in. At face value, time is the particles switching from one bubble to another, which results in a change of fourth dimensional frames in the five dimensional direction.

Chapter Two appears to be an ambiguous mess of rules that seem to be unrelated to one another. The point of Chapter Two as we have seen, was to lay the foundation for the universe in terms of laws. In order for us as to understand the universe, we must understand the most basic, and refined laws by which the universe was designed. For example, let us say that there was a child of infinite intelligence. The child would see that if he/she first realized the first postulate of the Laws of Fundamentals, by merely multiplying one of the values by negative one, he/she would have developed the idea of an equation. This is the utmost fundamental concept in math, but it gives rise to even greater understandings of the universe. Application of the rules set in Chapter Two, are the beginning of understanding of all things that reside in the universe, and it's existence.

One implication is the solution to the wave/particle duality. The truth in this matter is that the universe is subject to complexity, which allows nothing on a large scale to look as it does on the quantum scale. The solution is that, no matter what the universe appears to be, or look like at large scale, the universe is nothing but waves. Particles are the result of a macrological concept. We call a ball a particle, but in reality it is a collection of waves. If this is to be our definition of a particle, then particles are waves. Wave/particle duality exists, because we have not realized that everything is a wave. Cars, houses, thoughts, and so on, are all waves. The medium for such waves is energy, which is omnipresent, and when further investigated, we see does not exist by the inherent definition. We come to a well defined line when we talk about energy. It is the differentiating barrier between that which is considered to be, and not to be. All knowledge fails beyond that line. To make more than a handful of assumptions is to become absurdly speculative at best.

It should also be point out that the "luminiferous ether" does exist, and the idea was simply incomplete when it was thrown out. Moreover, the reason it was thrown out was, because the physics community at the time needed it only for travel of electromagnetic waves. There is no need for the ether in the travel of electromagnetic waves. This was proven by Michaelson and Morely with their famous interferometer. It was at this point that Einstein said, "I never said the luminiferous ether did not exist, I merely said it was not necessary in the travel of light." The luminiferous ether is a collection of spherical packets of space with a diameter of planck length. The packets of space-time get pulled to the center of a string that has been looped, and therefore has mass. When more than one string is present, the bubble will get tugged toward the midpoint of the two, and so on. By taking the sum of all the vectors, one could find the geometrically significant point of the mass, otherwise known as the center of mass. The reason an object has a center of mass is due to the bubbles converging at a point. Again, the rotation of the string is responsible for the moving of bubbles with respect to the object. The convergence of the bubbles is used today without realization. The significance of using vector addition in gravitational force problems is, because the vectors signify the net direction of the bubbles. These bubbles will also accelerate the masses involved in the same direction as the bubble, because the object exists in the bubbles that are changing position.

Another key concept deals with gravitons. There will never be a graviton found in any experiment, or in nature. They do not exist as they are currently defined. We will be discussing in depth the reasoning behind this in the following chapters. The other thing the author would like to mention is that this book will not refer to any force carriers. Particles of this nature are energy carriers, and there is only one in all cases. By observation, we see that no carriers are needed in gravity. This is also a foreshadowing for the Higgs particle, and field. There is much correlation yet to be uncovered.

The unsurprising thing is that the position of the bubbles is unknowable because in our experience, we perceive no such differentiation of space packets. It is not meant that we cannot see by experiment that these bubbles do not exist. As a matter of fact, we have proven that these bubbles are what the universe is composed of via relativistic Doppler shift, and gravitational lensing. Breaks in between the bubbles are where "nothing" exists, and these regions cannot be experienced, because of the absence of space-time, is outside the realm of experiment.

If the preceding chapters were correct, then the Law of Conservation will need to be adjusted, or completely thrown away. The author would not go as far as to say that the Law of Conservation should be thrown away at this time. Regardless, changes must be made to the law. The first proposed change deals with energy being a vector and not scalar. The proposed this change is:

The total amount of energy in a system is not constant. However, the average overall orientation of the string with respect to the 5th dimension must always decrease. Thus the amount of useful energy will decrease.

This adaptation will account for observances until experiment show otherwise.

Chapter 7
The Big Bang and Black Holes

The Big Bang theory was proposed in the early to mid 1900's. Before then it was thought that the universe was static. In fact Einstein had a hand in nearly everything that was going on in his time, and his equations caused it him to stumble across a great discovery. So revolutionary were his findings that he tried to adjust his equations with a cosmological constant to fit a universe static. It appeared that the universe was expanding. Edwin Hubble had first observed the expansion, and this was the birth of theory that changed the world's thinking. The Big Bang Theory suggests that the universe started as an infinitely dense ball of matter at the beginning of time. What happened before time began is uncertain, but at the beginning of time it appears that a ball of matter exploded out ward for reasons the theory has not yet explained. This resulted in the release of a quantity of energy that is known as planck energy, which heated the universe to a range of billions of degrees. When the Big Bang occurred, the universe underwent inflation. Inflation was a time when physicists suggest space-time expanded faster than light speed. The author does not agree with others in saying the universe expanded faster than light travels. The explanation of this period of inflation will be explained later. The Big Bang Theory also suggests that the universe is either still expanding, or contracting depending on the amount of matter in the universe, and what age the universe is. There are two possible endings to this theory. If there is a sufficient amount of matter, the universe will contract to a Big Crunch. If there is not enough matter in the universe, it should expand forever. The reason why the Big Bang even occurred has been puzzling since the discovery.

What the author has to propose would accurately answer many questions that otherwise cannot be answered. Instead of giving the reader the differences between the currently accepted theory, and the author's, we will go from beginning to the end of the Big Bang the way the Deftonic Theory suggests that it happened. It is widely accepted that the universe began with what is known as the (Hot) Big Bang. Supporting evidence is the background cosmic radiation that exists today, expansion (we know this because of the Doppler red shift), the amounts of particular heavy elements, the amount of Helium, and the universe is of a finite age.

The universe began as a void, a point of nonexistence. The Law of Fundamentals suggests that the universe would attempt to achieve equilibrium between existence, and nonexistence. The opposite of nonexistence is, of course, existence. For such a phenomenon to occur, certain parameters must be set. Remember time does not exist so we will need to use a form of time called regressive geometric time. This will at the very least keep a chronology of events so it will be less confusing, even though our chronology holds no weight in reality. So, in setting parameters for existence the universe needed a region to exist. Thus began the bending of dimensions. A point was bent into a one-dimensional line of planck length. This occurred at least four more times initially, which created the space-time needed to contain existence. At this point (time has not yet begun), the universe is in a cube shape, and gravity has not yet developed, as we know it. Now the intense instability of the system cause materialistic separation, and an undisclosed number of strings to split into two super dense spheres and stretch space-time. Before the two super dense bodies completely formed their respective loops of energy (matter), the Heisenberg Uncertainty Principle causes a massive release of energy. It did so, because these two super dense spheres were touching each other. So the particles would have to have been able observe the exact position, and velocity of one another. At 10^{-43} seconds otherwise known as plank time, the Big Bang had occurred. Formation of the sixth occurred at the same instant. This dimension's formation was caused by the lack of a unifying force capable of causing a universal collapse. Because of this, the universe took on a curled up shape. The sixth dimension did not form until after time began, thus changing the way the dimension formed (and whatever other dimensions may have formed). At the same instant the universe began to inflate. During inflation the universe did not expand faster than light. Actually, the entire universe underwent "travel" via an Einstein-Rosen bridge. This explains the large collections of matter, and the fairly random placement, or non-uniform distribution. If the universe expanded from the Bang to its current state without interruption, one would expect to find a more uniformly dispersed universe in terms of macroscopic arrangements such as super clusters. The individual galaxies would seem more so randomly distributed if this universal tunneling occurred, as opposed to the classical inflation. The inflation of the universe would defy Special Relativity if anything other than electromagnetic waves could travel at such a speed. It is absurd to think even in the early universe, that such a speed could be attained. If this concept is correct, and the universe underwent quantum tunneling on a large scale, then the universe should be considerably larger than we think. As a matter of fact, it seems very obvious that the universe has more than enough matter to recycle itself in a Big Crunch.

I would expect to find, for the most part, that the universe is shaped like a flat ring, or a spherical shell of matter with a nearly empty center. Depending on initial conditions of the universe, much of the light from across the other side of the universe may not have reached us. If it has, then there must be a considerable amount of Machos, Wimps, and Halos out there. Neutrinos are one of the particles confirmed to exist in recent years. Neutrinos fall into the category of Wimps, and were found in a large underground water tank. The neutrino was found in a tank some miles below the earth's surface that contained heavy water. The neutrinos would occasionally collide with the molecules, and cause little releases of light. The author expects much more mass to be in space than physicists theorize. The key to approximating closely the amount of matter in the universe could come from the study of ZPE. The only way such an approximation would come is if we were to calculate the maximum energy production in a completely unstable system. In saying production of energy, it is meant as stripping the universe's space-time continuum of untapped energy, and thus tapping it.

Let us assume that there is not enough matter in the universe for a Big Crunch for whatever reason. The energy death of the universe suggests expansion forever outward. Would not the universe become empty in particular spots, and unstable due to the absence of existence. This is the very foundation for the idea of Zero Point Energy, and materialistic separation. Within a finite period time (disregarding the large quantity of time, because we have "forever"), a universe below critical density would eventually achieve enough of a gravitational pull to draw everything back in on itself. This is the Boltzmann-Lorenz Law of Equilibriums, and Constant States. The truth, in any case, is that the universe will result in a Big Crunch.

In getting back to the cosmological constant, the author feels a mistake has been made. The author does not find it hard to believe that there is a cosmological constant. In fact, he would have to say, he would expect one. The fact of the matter is that the author does not believe this constant is due to any extraneous pressures as one may have heard in lectures. The definition of a force suggests that if and when there are two objects in the same orientation, and rotating in the same direction, that there shall be a repulsive force. Basic physics suggests that if a force is going to be experienced due to pressure, there will be a boundary to the system. As far as math suggests, the author sees no evidence of such a boundary in the universe. Yet observation may suggest that in some portions of space, matter is accelerating. When one looks at simple calculus and Riemann geometry, the first thing that comes to mind is positive, zero and negative curvatures. With that said let us looks at some math.

A. $\Lambda=(8\pi G/3c^2)p_\Lambda$ (energy density)

B. $\Omega=(8\pi G/3H^2c^2)p$ (density parameter)

These equations describe the universe in terms of space-time being a sheet, and forces being represented as curvatures. When Ω is less than one then the curvature of space is negative. If Ω is one then the space is flat. If Ω is more than one then the curvature of space is positive. Now, by looking at the Robertson-Walker metric we observe very much the same results within space-time.

C. $ds^2=c^2dt^2-a^2(t)[dx^2+f(x)^2(d\theta^2+\sin^2\theta d\varphi^2)]$

Where f(x) describes the spatial geometry of the universe parameterized by curvature constant. When $f(x)=[\sin\sqrt{(kx)}]/\sqrt{k}$, "k" then equals some value greater than zero the system is closed in terms of expansion. When f(x)=x, there is no k, the system is flat and susceptible to gravity alone. When $f(x)=(\sinh\sqrt{|k|}x)/\sqrt{|k|}$, and so "k" is less than zero, the system is open. It is difficult to tell exactly what is going on throughout space when we have such a small region to view from, although telescopes are becoming more and more powerful. Anyone interested in astronomy must keep in mind is that the light we get from the sun was emitted about eight minutes ago. Hence we are 8 light minutes away from the sun. This is interesting because the stars we look up at night show the history of the universe, because the light has taken so long to travel to us. What about the galaxies 100,000 light years away? We are actually looking at the past. So, how can we tell which way objects are doppler shifting right now? The answer, in short, for the cosmological constant question is not that there is some fifth fundamental force, or some pressure force. It is the compatibility, or incompatibility in some sort of wave behavior, depending on the curvature of space. The most likely candidate would be spinning and rotating objects causing such forces.

One of the latest ideas out is the thought that the cosmological constant is due to zero point energy shown in ht e following math;

A. $H^2 \equiv [(1/a)da/dt]^2 = (8\pi Gp)/3 + \lambda/3 - k/a^2$

B. $\Omega_{mo} \equiv (8\pi Gp_{mo})/3H_o^2$ and $\Omega_{\lambda o} \equiv \lambda/3H_o^2$ and $\Omega_{ko} \equiv -k/H_o^2$

And so the equation of motion becomes;

C. $(da/dt)^2=H_o^2[(\Omega_{mo}/a)+a^2\Omega_{\lambda o}+\Omega_{ko}]$

The author disagrees with these mathematics, because they are suggesting the possibility of such forces in miniscule regions, these forces would not have much of an effect on the universe on a macroscopic level. The problem with ZPE is that firstly, there are no such boundaries in the universe for which this claim has any validity. Secondly, the outer reaches of space (in terms of the bubbles) are not there until they are observed. The author is an advocate of the Copenhagen Interpretation. Being that these bubbles are to be considered particles made of an infinite amount of untapped energy, they, therefore, abide by quantum mechanical rules. Below is the mathematical version of the Copenhagen Interpretation:

A. $\tau=\int_\gamma\sqrt{[g_{ij}(dx^i/dt)(dx^j/dt)dt]}$

Basically we can cut down on the jargon by saying that the Copenhagen Interpretation has two main ideas:

B. Quantum particles have a definitive momentum and position.

C. Reality is "created" by the observer.

This coincides nicely with the string theory that the Deftonic Theory has put forth. There are strings that are quantized, and cannot make a complete circle, because pi is irrational. The particles do have a definitive momentum, and position. However, because of the previously mentioned circumstances, the particles are subject to exist when observed merely because the rules of nature disallow any such an observation. The result is a skewed amount of momentum, and position measurable at one time, thereby making the quantum world subject to critical observation.

When we look at the universe's "life" in terms of cycling, there is an overwhelming likelihood that the universe will result in a Big Crunch. The question is, "Where is all the mass?" There is speculation that Machos, which are massive bulks of dark matter, contain much of the missing mass required for a Big Crunch. The other projected source of dark matter comes from Wimps, in which there astronomical numbers of extremely low mass particles. The most prevalent idea that has been extrapolated is that nearly all the matter underwent an Einstein-Rosen Bridge. This further implies that the universe would be shaped like a ring, or spherical shell. It is possible that in such a case that we have not yet experienced the light from the other reaches of the universe either. The reason for this is, because the photons have not arrived, or the concentration of received photons is not observable with current technology. The supports this theory simply, because inflation says that the universe expanded faster than light travels. As mentioned, under no circumstances is the speed of light achievable by massive bodies. We cannot even today, accelerate a quark to light speed. Therefore, energy dissipation of the tunneling sort must have occurred.

It has recently been suggested that the speed of light may have been much greater in magnitude in earlier times. The founder of this theory is Joao Magueijo. This Varying Speed of Light Theory may be one of the strongest theories, which tries to better explain inflation. One ramification of this theory is that black holes cannot be entered into. The approaching matter never actually falls into the black hole. The current idea of black holes being "matter eaters" would have to be abandoned. Such abandonment would mean major disaster for much of the astrophysical concepts that are on the borderline of being accepted as physical phenomenon. The other most obvious problem is energy conservation in classical mechanics, and relativity. If "c" is changing, mass is constant (for non-relativistic systems), then what adaptation, if any, should be made to: $E=mc^2$

As it stands, if any mass is constant, then the Varying Speed of Light Theory becomes a victim of energy conservation. However, the major question posed by this book is, "Is energy always conserved, or is it possible under extremely odd conditions, that energy could be created?" To solve many problems in physics, which pose such a question, physicists have come up with an idea called negative energy. They call it negative energy, because it comes up as negative quantities on paper. The question now being posed is asking whether, or not this mathematical interpretation is physically occurring, or is energy being created?

The talk of Big Bang cannot go without the mention of General Relativity's most famous (infamous if you will) implication, black holes. Black holes are super dense stars (some argue particles) in which the gravitational force is so great that light cannot escape the field. There is so much known about black holes that we have uncovered in such a short time (less than a century). The one thing that should be known is that, even though black holes are widely accepted as a part of nature, it has not been directly proven that they exist. If the hard evidence should present itself, Dr. Hawking and a few other brilliant astrophysicists, would have a Nobel Prize coming to them. One of the greatest pieces of evidence, and probably first, was Cygnus X-1. Cygnus X-1 is believed to be a binary star system. The problem is that we see only one star, and yet it rotates around something. In such a case, one must infer that, whatever the star is rotating around is more massive than the star itself. Using the angular velocities of the bodies, the mass in question would have to be many times the mass of a neutron star. The suggestion is that this is a black hole.

Firstly, stars have different stages of their life. They may either be bright giants, or slightly luminous dwarfs (although some other categories apply). These stars are nothing more than gases, mostly light like hydrogen, and are constantly exploding outward via nuclear fusion. These stars are held together by gravitational forces that exceed any gravitational force we have ever experienced. In the case that the fuel runs out, the star explodes, and is referred to as a supernova. If a star is dense enough, it will become a neutron star. If the density goes much higher than that, a black hole forms.

The Big Bang ties in, because at the beginning of time, the universe supposedly began in an infinitely dense point called a singularity. Of course, the author does not agree with the prevailing thought that the universe started at a point, rather a bubble being the smallest volume, and therefore the smallest volume for a singularity. The question posed long ago dealt with the density of stars. If we crush down a giant star, it is classified as a dwarf. Further compression brings us to a neutron star, but what is next? What happens to space-time when you bend it so intensely? It is what we call a black hole. As we have already said, at certain distance from the center of mass, the gravitational force is so strong that light cannot escape it. In 1916 Schwarzchild worked out the equation for the Schwarzchild Radius, more commonly known as the event horizon. This imaginary radius is the point of no return in which absolutely nothing can escape. _Schwarzchild's equation is not as difficult to derive as it was just a matter knowing what equation to manipulate, and it so happened to be the escape velocity equation. The derivation was done earlier in this book, but the following is a slightly more detailed derivation:

A. $v=\sqrt{[(2GM)/r]}$
B. $c= \sqrt{[(2GM)/R_s]}$
C. $c^2=(2GM)/R_s$
D. $R_s=(2GM)/c^2$

This is a well thought extension of Newtonian physics, and this brings us to a big question. How did the Big Bang start as a black hole, and a singularity, and escape? This just supports Deftonic Theory. Deftonic Theory solves such problems by not being present at the beginning of time. The formation of the sixth dimension occurred the instant after the Big Bang. The escape would have been impossible without the entire universe undergoing quantum tunneling the instant before the formation of gravity. This is surely the most reveling postulation that has been suggested. When gravity was formed, most, or even all of the matter in the universe was already tunneling, and escaped gravity.

One of the greatest problems plaguing theorists today is whether there is such a thing as singularities at the "bottom" of black holes. All physicists have to go from is mathematics. There is a particular classification of black holes known as Kerr black holes and for which there are mathematically derived solutions as follows:

A. $p=b'/(8\pi Gc^{-2}r^{2})$

B. $\tau=b/(r-2(r-b)\Phi')/(8\pi Gc^{-4}r^{2})$

The next equation is for the presence of spin.

C. $p=(r/2)[(pc^{2}-\tau)\Phi'-\tau']$

In a static Scharzchild hole, Einstein-Rosen bridges are mathematically possible. A Scharzchild hole is classified by no charge, or spin. Reissner-Nordstron holes are black holes with a nonzero spin, or charge.

The other type of interesting black hole is the primordial black hole. This is a black hole caused by intense pressures, such as in the early universe. There is some speculation that we actually encountered one of these in the Soviet Union. It was when an unexplained force obliterated hundreds of miles of forest in Siberia. In fact there was a heavy disturbance on the other side of the planet as a large basin became energetic in terms of waves. These primordial black holes are expected to be anywhere from the size of a pinhead to a quarter, or larger like a softball. These types of black holes may be the proof physicists have needed, considering that going right up to a super massive black hole is not recommended. Possibly the only confirmation physicists may ever encounter is a laboratory produced primordial black hole, created by intense pressure.

Black holes are not only the result of extreme curvatures in space, but also the result of a buildup of energy in a region. There is no white hole for most black holes if any. The reason for this is, because there is no dissipation of energy. Since the energy is in the form of matter in the black hole, the energy does not get spent to form a wormhole._ This is the same principle procedure that Einstein-Rosen bridges, and quantum tunneling begin with. Stagnation within the universe is disallowed by the Boltzmann-Lorenz Law, which was proposed earlier. Therefore, there must be a loophole, and it turns out to be Hawking Radiation. These are all ways to distribute energy throughout space more evenly by dissipating it from regions of higher concentration. We have once more entered the realm of chaos and thermodynamics. The universe, despite its clumpy format, actually endorses even distribution, and at the same time ensures cosmological oscillation.

The most elating thought is that possibly there is a blue print in black holes that can explain why forces change at different temperatures. It is important begin figuring out what gravity was like at the beginning of time, and be able to explain how mere temperature changes the constants for the electric and gravitational field equation. The author believes that there is an answer very near, especially, because there has been extensive research done in this field.

In our talks of the ultimate abyss, we see a small piece of the beginning of the universe. What we call the Big Bang is heavily supported by observations of the universe today. This may be a good time to look over the evidence supporting the Big Bang Theory, which was stated earlier this chapter. The most important may be the wrinkles in time George Smoot, and his team uncovered in the early 90's. The universe ticks to the beat of a perpetual oscillation from Bang to Crunch. The universe is perfect in its design, and will never become stagnant by any means. The idea that went almost ignored in this chapter is the mechanics of gravity. This an inescapable force, even though it is the weakest. As we end this part of the discussion, it almost appears as though there is an underlying allusion in all of this is the goal of equilibrium. Black holes are a wonderful example of the Boltzmann-Lorenz Law, because the universe's end result would be a black hole, or a large region filled withy many black holes.

One thing that should be elaborated upon is Hawking Radiation. It was thought for some decades that black holes were black. In fact, the thought that light could not escape the gravitational field, lead to the name black hole. However, the universe has been constructed to behave in opposition to such stagnant states. Just outside of the Scharzchild radius is a nearly perfect vacuum. In this vacuum, a particle, and its anti-particle are created, and released. Three possibilities present themselves: immediate annihilation, matter particle escapes, or anti-particle escapes in to space while the opposing particle falls into the black hole. This gave way to the truth that most black holes are not entirely black. This is a very revolutionary thought that has changed the way we look at black holes.

The Big Bang can only be visualized through systematic adjustments in the theory, as more and more data is collected. While the Big Bang leaves only a few remnants of its occurrence behind, the evidence is insurmountable. Between background radiation, wrinkles in time, the amounts of heavy elements, the universe's expansion and so on, few other possibilities follow the criteria. Moreover, if there were any opposing evidence, it would take much to dissuade the physics community. The best possible source we have to explain those first instances are black holes. In studying black holes, we have found remarkable adaptations of design in the black holes for which the Boltzmann-Lorenz Law will be satisfied. A key to remember is that, when energy becomes extremely dense in regions, the universe has means by which that energy shall be more evenly dispersed throughout space. The problem is that most of the energy in a black hole is matter, and therefore cannot be dissipated. Hawking Radiation is that problem's answer. The very unification of General Relativity, and quantum mechanics lay at singularity.

Chapter 8

The Grand Unified Theory

There is much information to take in when dealing with the unification of forces, and fields. An in depth understanding of space-time is absolutely crucial. We started this discussion with what the forces act upon, then the forces of nature, and we will work them into fields. The currently accepted system or class of fundamental units of matter is called the Standard Model. This Standard Model is composed of six quarks, six leptons, and they interact via force carrier. There are four force carriers in this model at the present time. This Standard Model is seemingly simple at first glance, but then you step into particle physics, and chromodynamics. Here the classic laws of the universe appear to hold little, or no weight. The quarks and leptons, at this time, cannot be broken down any further, or so it would appear. Many physicists believe that they are the most fundamental particles in existence. The string theorists believe that quarks are strings dancing in space. It is not known to them whether these strings are looped or not. These theorists have more than few other questions to be answered. They are probably correct in their theory that the quarks are just waves of energy (strings).

These strings are definitely bound loops in the quarks, and the leptons are unbound loops. The mass confusion comes in leptons, which are low mass particles, and behave abnormally, such as electrons. Physicists are fairly sure that the electron is just a small field of energy. The author of this book concurs. According to the Deftonic Theory, leptons are loose, or unbound tapped energy. A visual image would look like a number of strings put into a spherical shape and it would look like a porcupine ball. The author is still unsure of the exact reason why these strings of energy do not just go everywhere. It is suspected that it has to do with the same phenomenon as the weak and strong force.

Let us start with quarks. There are up, down, top, bottom, strange and charm quark types. For each type of quark there are three flavors. They are blue, green, and red. This brings us to a total of eighteen quarks plus the anti-matter quarks, which raises the total to thirty-six quarks. Of course, if matter, and anti-matter meet there is almost always a large release of energy. Quarks can be put together to form mesons, and baryons. We are familiar with baryons, which are trios of fairly stable quark combinations to make particles, such as protons, and neutrons. Mesons are combinations of quarks that do not form stable particles. Mesons do form, but then undergo violent decays, and do so rapidly. The mesons are particles as such kions, and pions. These two groups of massive particles are called hadrons. The particles are bound by gluons, which we will get to later.

Leptons on the other hand are in separable. By saying inseparable, it is meant that if they were to be broken down they would not resemble their previous form. Leptons are nearly massless particles that misbehave the rules for which the hadrons obey. They are said to be point particles, but by the Deftonic Theory proposed, the author suggests otherwise. If a particles physicist were to try, and collide two of these leptons with one another they would find something interesting. No matter how fast someone was to send each lepton at one another, they would never collide thus giving a radius. In fact, if enough energy were put into the system, they would tunnel right through each other. At first sight the logical explanation is that they are point particles. If one were to apply the Deftonic's energy portion of theory, we see that leptons are loose, or unbound energy. Actually strings of energy are just bouncing around in the shape of a porcupine ball. When trying to measure the strings in the x-direction via x-directional collision, the strings have no width, and so the measurement of the y-directional string's width is taken. That is the whole catch, the width of energy is zero, and therefore these particles are taken to be point particles. In knowing that these leptons are unbound balls of energy, or fields of energy, they only differ in a few characteristics such as the amount of energy, and behavior of the field.

There are, or at least were, four Fundamental Forces some 50 years ago. They are the electromagnetic force, weak nuclear force, strong nuclear force and gravity. Weinberg, Salaam and Glashow united the electromagnetic force and the weak nuclear force to make them the electroweak force. The trio was awarded the noble prize for their accomplishment. So the race is on to unify the strong, electroweak and gravitational forces. This theory has earned the name the "Grand Unified Theory," or the "Unified Field Theory." Physicists have already begun unifying the strong, and weak force. The odd ball as it would seem is gravity. If we are going to first deal with forces, allow us to revert back to the definition of a force. In Chapter Two it was proposed that the definition of a force was: the compatibility or incompatibility of two wave functions. Knowing this we have already described gravity as the force caused by opposing ends of a bent rotating string, we shall attack the problem with a new perspective. In terms of the weak, and strong forces, I believe that these are the result of energy strings oscillating in different manners.

The author believes the strong force to be small yet intense field of oscillating strings. These strings almost completely fold over one way, and then fold back the other way. This results in a very short range force, because there is little disturbance in the surrounding space-time. The lack of disturbance in a large region of space is due to the amount of space-time interaction. Where the oscillating string is for the most part confined to a region of bubbles. The author believes gluons are the same occurrence of oscillating energy strings as in the weak force. Let's not forget that they can form "glue balls" and this happens, because there is no gravitational field for the strings to restore.

Particle physics tends to become abstract when dealing with these so called force carriers. This field of physics has a theorist group all to its own. There are supposed to be 8 gluons, 8 gluinos, 2 charged bosons and one uncharged, 3 winos, 2 photons (gamma and z), 2 photon zinos, Higgs particles (never to be found) and there are supposedly 9 of them in this family. As you can see, the fragmentations of string are named depending on the number of strings, and their behavior in those states. To me these are mostly details. The fluid diagram is the goal and will completely describe the particles in entirety.

If we look at particle physics and keep in mind the complexity of these systems, we take note that the same forces hold hadrons together. The difference is in the masses. The distinguishing factor between mesons, and baryons is the ability to contain all of the particles with in the field. It takes two up quarks, and a down to make a proton. This brings the total mass to approximately 1080 MeV. This is a stature of complexity for which the gluons can maintain the compact group of quarks. Remember, the gluons are not in control once a particle leaves the region of the field in terms of attraction. In fact, there is little, or no force beyond that point, thus setting parameters for the maximum mass of particles, and the amount of time (on average) for breakdown. If we were to throw together randomly three quarks we get the mass of about 3540 MeV. Such can be found in experimentally to decay in a fraction of a second, because of the succession of the gluon field.

The weak force is very much the same as the strong force, but with one major difference. The weak force only works in half oscillations, and because of that is a slightly repulsive force. Moreover, it will aid in the formation of new particles. The minimal interaction with space-time will result a short range force. The electromagnetic force is much like the weak force in the way the strings of energy curve. The cause of electromagnetism is string doing a sort of dance in space. Imagine a looped string giving itself some slack, and either collapsing inward, or jutting outward at points. That is all that the electric force is. At this point the author cannot tell whether the collapsing type of string is what we call a negative charge, or a positive charge. But these inward, and outward oscillations result in curvatures in space. These are long range forces, because they are highly interactive with the bubbles around them. There is such a correlation between gravity, and electromagnetism, because they are types of oscillation strings can experience, and unlike the strong, and weak force, they interact with the space around them to a very high degree. The unification of the forces is not the difficult part. The hard part is figuring out what temperature has to do with it all, and it may come down to how space-time reacts to these oscillations.

The last things we need to talk about before going on to fields are the force carriers. There are three force carriers, and one theoretical force carrier in the Standard Model. There photons for the electromagnetic force, W and Z bosons for the weak force and gluons for the strong force. The suggested theoretical particle is the graviton for gravity, which will be shown not to exist. It cannot be said that the graviton does not exist completely, because there is a such entity attributed to the job except the particle known as a graviton, as particle physicists talk about it, does not exist. Their description is most likely incorrect. In fact, the graviton is the same thing as the Higgs Boson. The first thing that we need to see is that these so called force carriers, are just energy traveling from place to place, and behave a certain way in a local region. The photon, as we have already gone over, is a single string of energy traveling through space, and changes its geometry. W and Z bosons are nothing more than photons trying to fully loop, and either disturbs quarks to change them into another type of quark, or do a few other little tasks. Gluons are oscillating photons, or completely oscillating bosons. The fluctuations in this type of energy field tightly bind quarks, protons, neutrons and so on. If you can name it, and it is looped energy it will get stuck in this energy field.

To explain gravity, a man by the name of Higgs suggested a field theory in which the "heavy" objects interact with the field more so than "lighter" objects do. We can imagine it as an evening soiree in which people are evenly distributed through out the room. When an important, or well-liked person comes into the room, the surrounding people gather around, and bunch up. For "force carriers" we can imagine a rumor traversing the room, hence there is little clumping of people and little interaction with the field. The field is known as the Higgs Field.

The theory goes further to suggest that there will be a Higgs particle, but not much is known of this Higgs particle.

Higgs was very much on the right track. There was much over looked though. The Higgs field is actually the quantized packets of space-time that has been called bubbles. They are one in the same if one were to think about it. The looped rotating string causes bubbles to change relative positions with respect to the string, and unify at the geometrically important center of mass. Where the collective vector sum of multiple loops of string is zero is the center of mass. The more strings a particle is composed of, the more it interacts with space-time. We can almost imagine the loops of string eating the space from all directions. The Higgs particles that are being talked about are the bubbles themselves. With further theoretical investigation you will see that these bubbles of space-time do meet the criteria to be considered particles. They move relative to other particles, fall subject to Heisenberg Uncertainty, and are completely composed of (untapped) energy. This is why the author believes this particle would otherwise never be found. It is not a matter of knowing where to look, rather it is a matter of knowing what you are looking for.

As has been said before, the graviton and the Higgs boson are the same thing. In fact they tie together all of the other force carriers. Gravitational potential in objects is held through a chain of bubbles that constitute space, and are constantly being drawn or accelerated due to the rotation of looped string. These bubbles are the force carriers in all situations. The mere fact that the transfer of energy most of the time carries a couple of descriptive characteristics especially mass, charge and spin. Regardless, these carriers are the space-time continuum of quantized packets of space, and the wave like motion thereof. Now there seems to be a challenge as far as Michaelson, and Morely go. It is not the mere fact that there is an ether that allows photons to travel. This is a certain fallacy. The truth is that back in the early 1900's the physicists did not know that space-time was getting "sucked" in toward masses, and unifying the bubbles at a geometric location known as the center of mass. For all they knew, the ether was completely "stationary." Moreover, the photon should really be looked at as not only the particle, but also a very simple wave. Imagine a pebble falling into a placid pond. The pebble striking the water describes the atomic activity, and the waves are energy, which in their wave format are bent and so have mass. The differentiation in height of the waves is the carrier otherwise known as space-time (bubbles). All forces are like this in their behavior, and it does not matter whether they are long range, or not. To get off the subject of fundamental forces for just a second, think about Jack who is going to push a box. The incompatibility of the box's and Jacks wave function cause a curvature in space. Once the push is relinquished, the curvature in space is alleviated. If there were long range incompatibilities to entice the space between the objects to bend, then there shall be a force present. (The actual incompatibility is due to the electric force between electrons in the atoms.)

Now we move on to fields, which can be very tricky. Let us first make sure we know what a field is. Field is short for force field. All that a field is, is the geometry of the space-time around the source of force. Einstein spent the last years of his life in a hunt for the monster itself. It was too elusive at that time for even the greatest mind of the 1900's to capture. But he made much progress and left an important equation behind. Each force field for each fundamental force is somewhat different. The force intensity at different points in space is what defines the field equation. The electric field equation, and gravitational equations for force at different locations in space, differs only in a constant. The actual field equation that Einstein came up with for General Relativity was as follows;

A. $R_{ij}-(1/2)Rg_{ij}-\lambda g_{ij}=[(8G\pi)/c^4]T_{ij}$

At the time Einstein worked on the unified field theory, and many of his peers felt that he was wasting his time. Today there is a resounding difference in opinion with in the physics community. The main thing now is that all of the forces that we have tied together in terms of space-time curvatures, and their causes. In delving into the mathematics we may need another Issac Newton for the 21st century to unify the fields mathematically. The math is there, but no one can find the solutions leading even into the proper direction. The difference in the fields' behavior is definitely related to the temperature of the system for whatever reason. If the reason behind the differentiating of the fundamental forces at differing temperatures can be found, the math will become much easier. As it is, the author has provided the theory for which the theories of relativity and quantum mechanics have been unified in a self consistent space-time theory. It is truly a pinnacle pending upon its acceptance by the physics community.

In the cases of the electric, and gravitational fields, there are a long range forces, the motion of the bubbles are either proportional to the force experienced at ascending radii, or inversely related. For example, the gravitational field has only one behavior in which the bubbles can "move." Due to the fact that gravity is the most unique force, (each particle has both positive and negative charge) the force is always proportional to the force at the varying radii. In the case of the electric field, the charges are separated between bodies of particles. The author does not know whether negative charges are either kicking out bubbles of space, or taking them in.

A suggested experiment to prove the relative behaviors of the electric field, and gravitational field, is to create a massive charge. Then find whether time is running more slowly toward the center of the one type of charge, and then finding whether the converse is true with the opposite charge. It comes down the General Theory of Relativity. If Deftonic Theory is correct, then General Relativity should apply to charge as well as mass. Moreover, this experiment will tell whether positive, or negative charges are taking in bubbles.

Chapter 9
Final Thoughts

In the preceding chapters we have covered much groundbreaking research and thoughts, and to the best of the author's knowledge is completely consistent with the observations currently being made. We lead things off with a very interesting view of time and tied it to distance. Then we unified the two quantizations of distance, and time by saying that they are the same, and in doing so we ended up finding out that is the physical distance between perpenular sheets. We can imagine the universe being a giant cartoon flip book with a distance between the pages called planck distance. This minimum distance between pages also leaves us with planck time. We arrived at all of this by a simple process of bending dimensions, and that idea is called the flatlander idea. Of course, being dimensions of the space continuum, these perpenular sheets can be elongated, or stretched otherwise known as time dialation.

The quintessential idea of the universe is that all that resides within the system are waves, regardless if they are macrologically known as particles. This is called the Universal

Principle of Waves. We also went over the types of behavior. The most important item in dealing with waves is the Law of Fundamentals giving not only the structure of the universe, but also the inherent and simplest interests of the universe. That urge is to achieve equilibrium with whatever has been innately defined to be the opposite, and oscillate. The problem with this state of equilibrium is in the stagnancy of the system, the Boltzmann-Lorenz Law disallows a steady state. This sort of death, otherwise known as equilibrium, in the system disallows any such further oscillations, and therefore is remedied. In all truth there is no state of equilibrium that will remain as such. No matter what we do, or what the circumstance, the universe is symmetric in everyway, and will oscillate.

The better part of this book has been spent on energy. The energy that makes up the universe's fabric, and in the form that gives us matter, and tapped energy such as light. Energy is nothing more than one-dimensional lines of set quantized lengths. The Law of Fundamentals gives us a somewhat ironic view of the universe. The law says that for all that exists, there is an equal and opposite set of fundamentals that constitute it. The irony is that energy has no equal and opposite. Therefore, it does not exist and yet we have the universe that lays before us. The author finds it somewhat disturbing that due to the persistence of a law we have something from nothing. However, the age old question was how did we get this something from nothing. Between the Law of Fundamentals, and the Boltzmann-Lorenz Law, the universe has found means for which such not only could occur, but must occur. With such laws present, the universe has no choice but to have a temporarily existing component. Everything in the universe cancels to zero with the exception of energy, so no laws are broken. Energy does not even matter considering it does not exist, and that is why we need not worry of it. No one has ever encounter or observed energy. No one has ever calculated the total amount of energy in the universe. As physicists we must realized that we do not see the energy itself, rather we see the interactions of two forms of energy. We can only calculate the amount of energy exchanged in the system.

Forces are what keep the universe churning. The Universal Principle of Waves says that all that is observable, or exists in our system is a wave. The universe seems to unveil this more, and more everyday as we pay more attention. The search for symmetries in math, and the universe give us a feeling that there is sort of divine split down the universe. This thought is the father of force. The idea that in specific, forces have components that display compatibilities, and incompatibilities.

The problem that exists today is the lack of thinking of the universe as one system of one quintessential building block. The relay partner of forces, and carriers is the space-time itself. The idea of space-time that is a force carriers was long ago dismissed, and for the wrong reasons. An ether that is "eaten" by mass due to forces of gravity would have turned some heads back in the day. The author stands firm in the particle idea of the ether for which we have referred to as bubbles for so long. The only problem with the Michaelson/Morely experiment with the interferometer in 1887 was that they made the assumption that photons were particles that travel from point to point in space as we thing of a ball through space. There was a small error in that the photon is subject to particular types of geometric curvatures in space. However, the mistake made was in how the photon travels. The universe vibrates as objects travel, and this was not taken into account. The bubbles of space-time undulate at their maximum frequency as the bubble to bubble travel occurs. By means of wrong assumptions, the ether theory was discarded. The truth was that no medium was necessary for travel, but this is not the case.

Many theories have been introduced in this book, and they are all related in some way, or another. Many may seem like either a revelation, or an off the wall way of thinking. The author is confident that the work here is for the most part correct. Overall, scientific thought is a search for the truth. Since the early philosophers to the physicists of today, one could simply notice that ingenuity in itself gives rise to these truths. Ingenuity may come in several shapes, and forms. It may be a happy thought, an accident or an entire revolution in a wave of new thinking. There is one truth in all of this and that is that such ideas come at a cost. Whether it is the restructuring of a modern physical model, or just the ability to swallow down something that is barely understandable, and yet holds an unquestionable integrity. Genius comes at a toll called acceptance. The acceptability in an idea is the fine line that borders the university and the asylum and the genius from a crackpot. A line that is measured in success for which is unachievable without acceptance in the community of physics.

Being a young theoretical physicist, and astrophysicists may be the best perspective of the world one can have. A world still somewhat familiar and yet there is an intense enthusiasm inherent in the mind. A young mind has not yet been corrupted by the textbook style of thought that tends to steal so much ingenuity from physicists. A sense of hunger for truth and the doctrines of how to learn about the world have not yet been fully thrust upon such an individual. This has given the author ability to go about solving problems in such a way, and therefore, developing new and somewhat radical methods. Moreover, giving the author an inherent intuition over the resolution of problems that have not been answered. The carries on these intellectual processes in a manner which breeds to ingenuity. For the mathematician, solutions reside on repetition, but for a physicist, the ability to solve problems comes only with his/her capability of looking at the universe from a new perspective.

Perhaps in seeing the universe from a new perspective, the author has answered problems that have plagued the greatest physicists of our time or at least has begun to solve them. These theories may bring about the beginning of a revolution in all fields of physics. One of the most important things to keep in mind is not only whether the Deftonic Theory is completely right, or not. The most important things are the direction that the theory is taking, and the original perspectives that this book takes on. However, with accuracy that the Deftonic Theory solves so many problems, it is hard to image it being completely false. With that said, the Deftonic Theory will at least be partially accepted, if not completely, and it is only a matter of time.

Theories

1. Perpenulus is the 5th dimension, which is a flat sheet that represents the three-dimensional world from all angles at once.
2. The Fifth dimension makes energy a vector and gives perpenulus the ability to change, also known as time.
3. Law of Fundamentals
    a. There is an equal and opposite for all that exists.
    b. All fundamentals have an inherent nature to achieve equilibrium.
4. Rules of Wave Behavior
    a. Assertive
    b. Paradoxical
    c. Equilibrial
5. The Universal Principle of Waves; All that is observed within the system or exists is a wave.
6. Milliern's Principle-As entropy in a system increases, the complexity of the system increases until a breakdown occurs.

7. Phi and pi are structural constants; The severity of importance in phi is due to the fact that phi suggests that nature continually adjusts parameters of the system to better suit a higher complexity.
8. Interconnectedness Conjecture-The universe can be represented as one large multidimensional wave for which all things in reality are related.
9. Definition of a force-the compatibility or incompatibility in wave functions.
10. Boltzmann-Lorenz Law of Equilibriums-No system in the state of equilibrium or any other state shall remain stagnant in that state.
11. Energy is a one-dimensional string.
12. Perpenular friction results in dragging of space-time associated in Special Relativity.
13. Gravity is the result of strings of energy spinning and then bending.
14. Regressive geometry-Most importantly that all positions in space are the same location.
15. Space-time comes in quantized packets I call bubbles.
16. Materialistic Separation purges the instability of empty space.

17. Quantizations of length, energy and time are the same thing.
18. The sixth dimension appeared the instant after the Big Bang and curled up because the universe was already expanding and gave rise to gravity.
19. The ether is the Higgs field and the graviton is the Higgs particle, which I call a bubble.
20. Heisenberg Uncertainty is due to the fact that quantized units of energy cannot completely form structures based on phi and pi such as a circle.
21. The energy dissipation system of the universe is quantum tunneling, Einstein-Rosen bridges (wormholes) and black holes.
22. The early universe's inflation was actually the quantum tunneling of the universe.
23. Movement of bubbles begins to define the unified field theory.
24. Time, energy, and distance are the same entity.

Contributors

*Sir Issac Newton

*Nikola Tesla

*Albert Einstein

*Richard Feynman

*Niels Bohr

*Stephen Hawking

* Werner Heisenberg

*Edward Witten

*Louis DeBroglie

Edwin A. Abbott

John Bell

David Bohm

Ludwig Boltzmann

Max Born

Satyendra Nath Bose

Casimir

Arthur Compton

Paul Dirac

Doppler

Sir Arthur Eddington

Leonhard Euler

Michael Faraday

Leonardo Pisano

Galileo Galilei

Carl Gauss

Sheldon Glashow

Brian Green

Nick Hebert

Peter Higgs

Edwin Hubble

Christian Huygen

Michio Kaku

Theodor Kaluza

Klein

Thomas Lake

Gottfried Leibniz

Heinrich Lorentz

Edward Lorenz

Benoit Mandelbrot

James Maxwell

Albert Michaelson

Mills

Hermann Minkowski

Edward Morely

Wolfgang Pauli

Roger Penrose

Max Planck

Eugene Podklentov

Podolsky

Martin Rees

Georg Bernard Reimann

Rosen

Abdus Salaam

Erwin Schrodinger

Schwarzchild

George Smoot

MJ Sparnaay

Stefan

Steven Weinberg

John Wheeler

Yang

Thomas Young

Yakov Zel'Dovich

Zeno

*—Denotes major contribution

Glossary

Assertive Wave Behavior- Attempting to achieve equilibrium when there are two extremes, and no medium;

Boltzmann-Lorenz Law- No steady state including equilibrium will remain in such a state for an infinite amount of time;

Complexity- The potential for breakdown;

Constituents of the Void- Subset of the universe that lacks the parameters for existence;

Deftonic Theory- the theory explaining higher dimensional theory, and quantum mechanics by unifying them into onc entity;

Energy- One dimensional string that is the fabric of the universe, and all that it contains;

Equilibrial Wave Behavior- Attempting to achieve equilibrium when there are two extremes, and a medium;

Force- The attraction, or repulsion of two entities either by compatibility, or incompatibility of wave functions respectively;

Law of Fundamentals- For all that exists, there is an equal and opposite;

Milliern's Principle- This law states that the complexity of a system may increase, or decrease, but entropy shall always increase;

Paradoxical Wave Behavior- Attempting to achieve equilibrium when there are two extremes, and no medium by existing at two extremes simultaneously;

Perpenulus- Four dimensional space;

Perpenular frame- Theoretical "snap shot" of perpenulus without change in the fifth dimension;

Perpenular friction- The tugging of space-time caused by particles larger than each individual sixth dimensional bubble of quantized space;

Universal Principle of Waves- All that resides in the universe is a wave, or can be broken down to be a wave;

Void- Zero dimensional region that does not contain the parameters to harbor existence;

About the Author

David William Milliern was born on September 12, 1982. At a youthful age, David gained an interest in mathematics, but even before then, he found himself amused with the mechanics of the physical world. David began working through mathematical proofs and thought experiments during his years in public schooling, and has continued his studies at the University of Pittsburgh. David is currently a prominent undergraduate studying Astronomy, Physics, and Mathematics.

His interest in a dozen major topics in theoretical mathematics, physics, and astrophysics has brought about the book, "A Matter of Time." David aspires to become a tenured professor of astrophysics once his schooling is complete.

www.ingramcontent.com/pod-product-compliance
Ingram Content Group UK Ltd.
Pitfield, Milton Keynes, MK11 3LW, UK
UKHW041944190726
13854UKWH00004B/1784